COLUMBUS

An Annotated Guide to the Scholarship

on His Life and Writings, 1750 to 1988

COLUMBUS

An Annotated Guide to the Scholarship on His Life and Writings, 1750 to 1988

by
FOSTER PROVOST
Duquesne University

Published for
THE JOHN CARTER BROWN LIBRARY
by
OMNIGRAPHICS, INC.
PENOBSCOT BUILDING, DETROIT, MI 48226

This book has been endorsed by the United States Christopher Columbus Quincentenary Jubilee Commission as an Official Quincentenary Project.

This work is not to be reproduced in any form, in part or whole, for any purpose, without permission. Correspondence should be directed to the John Carter Brown Library, Box 1894, Providence, Rhode Island 02912, or to Omnigraphics, Inc., Penobscot Building, Detroit, Michigan 48226.

Library of Congress Cataloging-in-Publication Data
Provost, Foster.
    Columbus: an annotated guide to the scholarship on his life and writings, 1750 to 1988 / by Foster Provost.
       p.   cm.
    Includes indexes.
    ISBN 1-55888-157-3 (lib. bdg.: alk. paper)
    1. Columbus, Christopher—Bibliography. 2. America—Discovery and exploration—Spanish—Bibliography. I. John Carter Brown Library. II. Title.
Z8187.P76 1991
[E111]
016.97001'5—dc20   90-27572
                                                           CIP

The John Carter Brown Library is an independently funded and administered center for advanced research in the humanities at Brown University.

Typesetting by Verbatim, Inc., Providence, Rhode Island.

∞
Printed on acid-free paper, meeting the ANSI Z39.48 Standard. The infinity symbol that appears above indicates that the paper in this book meets that standard.

Printed in the United States of America.

For
Waldo F. McNeir

Table of Contents

Foreword

The planet Earth has become in the twentieth century a single, interdependent whole. This unification and integration began at a precise moment in 1492 when permanent contact was first made between inhabitants of the two halves of the globe who hitherto had been totally unknown to each other and, in fact, totally unimagined. Such an event could happen only once in human history, and despite its many tragic consequences, is cause for acclaim.

Our prideful hindsight leads us to assume that this meeting of completely separate and insular worlds was somehow "inevitable" and that if Christopher Columbus had not launched his venture into the unknown in 1492, it would have happened anyway—in 1496, or 1500, or 1510. Perhaps so. Yet a case can be made that it is only in the retrospective, creative act of writing history that the sequence of events in the past becomes inevitable; history as lived, today and five hundred years ago, is always astonishingly contingent and unpredictable.

Because of this contingency or freedom, the actions of individual men and women can create vast shifts in the orientation of humanity, earthquakes that change the landscape of history permanently. It was only after, and largely because, Columbus had the extraordinary intrepidity to sail west into the uncharted ocean, and survived, that scores of other voyagers followed. It was Columbus's exploit that made thinkable, within only a few decades, the Magellan expedition that circumnavigated the globe.

We can only speculate how and when contact between Europe and the Western Hemisphere might eventually have occurred, had Columbus not sailed into the unknown in 1492 expecting to find China or Japan. It is unlikely that the initiative for attempting to cross the oceans would have come from the Americas, and Europe in the sixteenth century, lacking Columbus's initial stimulus and feeling complacent about the route to the East around the Cape of Good Hope, could have become deliberately insular. Many decades could have passed without the opening of contact between the Western Hemisphere and Asia, Africa, and Europe, with its aftermath of fateful consequences for the history of humankind. Nevertheless, at a particular contingent moment in 1492, Columbus, a person with no striking particularities that destined him for this role, did the unprecedented deed, and made the voyage.

It is because of the fortuity of the "Discovery" that Columbus will always be a magnet for human speculation, reflection, argument, debunking, and emulation. Despite our awareness of the larger factors and forces that govern human events, and despite scholarly disparagement of the role of individual decision and resolve in the unfolding of history, inescapably we come back to Columbus because he is one of that small number of individuals who by some odd chance take those unprecedented steps that put him or her at the very center of epoch change.

Professor Foster Provost was a Research Fellow at the John Carter Brown Library in 1984 when he first indicated to me that he intended to commit the remainder of his active scholarly life to the study of Christopher Columbus, in the widest sense. This commitment has included ultimately not only the preparation of the bibliography that follows, a daunting achievement, but also the composition of an epic poem about Columbus—a small portion of which the Library has published, *Columbus: Dream and Act. A Tragic Suite* (Providence, 1986)—and even the editing of a semi-annual newsletter (also published by the Library) that covers scholarly events related to the Quincentenary, *1992: A Columbus Newsletter* (1984 –   ).

For his manifold activities concerning the life of the great navigator and the scholarly treatment of that life, Professor Provost needed an institutional base, which the Library was privileged to provide. We did not know in 1984 whether Professor Provost, or anyone, could prepare an annotated guide to modern scholarly writings about Columbus, so vast is the literature and so international its scope. Nor did we have a publisher or the funds in hand to sponsor the work. But Professor Provost had behind him a record of accomplishment—notably the compilation of *Edmund Spenser: An Annotated Bibliography, 1937-1972* (New Jersey: Humanities Press, 1975) prepared with Waldo F. McNeir— and we optimistically agreed to collaborate, assuming that if the world did not yet know that the greatest public commemoration in the history of mankind was coming along in 1992, it soon would.

The John Carter Brown Library, of course, had its own obligations to 1992. It is the only library in the world dedicated solely to collecting and preserving primary sources related to the discovery, exploration, settlement, and early development of the Americas, from Hudson Bay to

Patagonia, including both European and American sources. The very foundation stones of the Library are various editions of Columbus's 1493 "letter" to the Spanish court reporting on his earth-uniting voyage. The existence of the John Carter Brown Library since its founding in 1846 is predicated on the belief that the European discovery of the Americas in 1492 was the greatest secular event in human history, a unique occurrence that lies at the base of the modern world and that the colonial period, the period of European domination of New World affairs, would always deserve scholarly study and reflection—study and reflection not only on what happened in America as a result of the encounter and conquest but also what happened in Europe and Africa. Clearly, the mission of the Library and the personal commitment of Professor Provost were mutually enhancing.

Among other things, the Library had already made a start on a special kind of Columbus bibliography, its European Americana project, the research for which had begun in the 1970s. *European Americana: A Chronological Guide to Works Printed in Europe Relating to the Americas, 1493-1750*, 6 vols. (1980-   ) by its nature includes among its 30,000 entries virtually every printed reference to Columbus before 1750. Professor Provost's *Guide* may be thought of, in one sense, as picking up a strand of *European Americana*, studies on Columbus the man, and carrying it forward from 1750 into the late twentieth century. Thus, the Library has been given the opportunity on the occasion of the Fifth Centenary of the Discovery to provide scholarly tools for the study of Columbus covering nearly five hundred years of publication.

It is a pleasure to acknowledge here that for the Provost *Guide* we were helped along the way by a generous grant from the Culpeper Foundation and by the personal generosity at a critical moment of Dr. Peter Sammartino. Later, Mr. Frederick G. Ruffner, Jr., the president of Omnigraphics, Inc., emerged as the crucial underwriter of the European research needed for the bibliography. On the editorial side, the project has benefited from comments on the text by Professor Gabriella Airaldi in Genoa. For this assistance, we are most grateful.

Norman Fiering
John Carter Brown Library

Preface

My prefatory remarks to this *Guide* consist of thanks and acknowledgments. In the Introduction that follows I describe the fundamental assistance of Dr. Martin Torodash and of my wife, Rina Ferrarelli Provost. Other supporters are legion, first among them Dr. Norman Fiering, Director and Librarian of the John Carter Brown Library, who commissioned the work, helped generously in raising funds to support the commission, and supervised the typesetting of the book. Mr. Frederick G. Ruffner, Jr. of Omnigraphics, Inc., besides taking the publication under his wing (in cooperation with Dr. Fiering and the JCB Library) authorized the grant that enabled my wife and me to do essential research in Genoa and Madrid. Senator Paolo Emilio Taviani, chairman of the Italian Columbus Quincentennial Commission, has aided the project in many essential and generous ways.

The following persons also provided firm support: Dr. Robert Tolf of the Phileas Society; Dr. Juan Pérez de Tudela Bueso of the Real Academia de la Historia, Madrid; Hon. Frank Tumminia, Consul General of the American Consulate in Genoa; Hon. Michael Hahn, Consul for Information and Cultural Affairs at the Genoese Consulate, and Dr. Anna Maria Saiano and Ms. Margherita Mazza, his assistants; Dr. Laura Malfatto, curator of the distinguished Columbus collection of the Berio Civic Library in Genoa; Dr. Christian Zacher of Ohio State; Dr. Dennis Landis, editor of the *European Americana* project at the John Carter Brown Library; Mr. Josiah Marvel of Providenciales, Turks and Caicos Islands; and my expert readers, Dr. Wilcomb E. Washburn, Director of the Office of American Studies at the Smithsonian Institution; Dr. Delno C. West of Northern Arizona University; Rear Admiral William Lemos, USN (Ret.); Dr. Julian Granberry of Horseshoe Beach FL; and Mr. Kirkpatrick Sale of New York City.

I also received important assistance from Dr. Harrison Meserole of Penn State and Texas A. & M.; Dr. Michael Gannon of the University of Florida, Gainesville; Dr. Theodore Beardsley, Jr., Director of the Hispanic Society of America; and Drs. Michael Weber, Paul Pugliese, Jack Hausser,

Wallace Watson, Joseph J. Keenan, Albert C. Labriola, Carla Lucente, Ruth Hicks, John Hanes, and Coleman Myron of Duquesne University.

A word on library personnel. I owe a special word of praise and admiration to the personnel of the many libraries that I have used in the United States. They, as much as any group I know of, are selflessly dedicated to the dissemination of learning. They are generally underpaid and they work long hours at frequently inconvenient times; yet nothing is clearer than that the library profession in this country has shaken free of the old tradition of dragonship in which the ideal is to keep all the books safe and unused on the shelves.

Specific librarians and library personnel to whom this project is indebted, besides those already mentioned above, include Ms. Susan Danforth, Dr. Everett Wilkie, Dr. Ross Dealy, Mr. Daniel Slive, Ms. Karen de Maria, Ms. Lynne Harrell, Ms. Marie Martins, and Ms. Vivian Tetrault of the John Carter Brown Library, as well as the entire staff of that distinguished institution; Mr. Harry Hutchinson and Ms. Patricia O'Kane of the Duquesne University Library; and Ms. Kim S. Perry and Ms. Mary Lou Cummings of the Knights of Columbus Library in New Haven, Connecticut.

Foster Provost
Duquesne University
and the John Carter Brown Library

Introduction

This *Guide* presents a short but highly articulated history of scholarship on Christopher Columbus, in the form of an annotated bibliography arranged under appropriate topics. The list of entries is radically selective: it contains just under 800 items drawn from many thousands of books and articles published on Columbus in the period covered, 1750-1988.

Arrangement

The arrangement is simple: a categorized chronological sequence. First come collections of documents, texts, and studies (Chapter I), then editions and transcriptions of the primary documents (II) and studies of these documents as documents (III). Then come studies of Columbus's life either as a whole or in part (IV); then studies of "Columbiana" (V), i.e., of those various matters like Columbus's ships and his place of burial that lend themselves to separate study divorced from various other matters. After these come bibliographies (VI) and, finally, the historiography of the subject (VII).

The items in each section appear chronologically by year to enable the user to follow the historical evolution of the scholarship wherever possible. Within any particular year the items appear alphabetically by author except where the user's understanding can be enhanced by placing related items in another sequence: *e.g.*, the entries for the reviews deemed appropriate for inclusion as individual items come immediately after the book reviewed unless something about the review calls for its placement elsewhere, as with reviews dealing with more than one book, *e.g.*, #95. In the final chapter (VII), several review articles surveying Columbus scholarship are listed (#s 759, 765, 769, 772, 773, 775, 776, 777). Various annotations scattered through the *Guide* mention still other reviews, although there is no attempt to provide an extensive record of reviews.

Annotations

Each item in this *Guide* is annotated. The policy I have adopted for annotation is to report on each item objectively without prejudicing the reader as to its value. My responsibility for maintaining a critical approach was discharged initially in the act of selecting items for inclusion: the presence of an item in this compilation indicates my judgment that the item is worth examining. Nonetheless, I have not always felt it appropriate to exclude comment on the quality of the item being annotated. Sometimes the influence exercised by a particular item requires its inclusion even though the scholarship is not sound, as with Roselly de Lorgues's 1856 book *Christophe Colomb* (#655). In this instance the annotation asserts my opinion that the author has indulged in unjustifiable fictionalizing, as in his account of Columbus's mistress, Beatriz Enríquez de Harana.

In my continual attempt to place items in context by cross-references to other items in the *Guide*, I have tried to keep the entire body of Columbus scholarship in mind. In this connection an occasional *caveat* appears in the annotations where a warning has seemed appropriate. A chief reason for including some items requiring such a warning (instead of merely discarding them) is that occasionally an item whose scholarship is seriously defective (like Roselly de Lorgues's biography, #655) suggests ideas that establish the course of more competent subsequent scholarship or (as with Roselly's book) irritates better scholars enough to inspire more thorough or more accurate treatments (see #656). Also, when a particular area of scholarship contains few notable studies or none, the need for further study can sometimes be highlighted by citing one or more less-than-adequate studies in the area (see #609).

With few exceptions, the annotations summarize the item or indicate its scope and purpose. Where possible, especially with complicated material like that treated by Edmundo O'Gorman (#s 535-538), the annotation reports the argument in some detail. I have personally examined almost all of the items listed here and have personally annotated almost 90 percent of them. For the rest, I have depended on annotations made by reliable scholars who had seen the items. Wherever I myself have not made the annotation, I have cited the person who did make it. My chief sources for the annotations that I did not make myself have been Martin Torodash, cited below under "Sources," Donald H. Mugridge (#735), Charles E. Nowell (#765), and my wife Rina Ferrarelli Provost, who worked with me on this project in Genoa and Madrid for six weeks in the summer of 1989. A few other annotations, which I have translated and sometimes abstracted, appeared in the annual annotated bibliographies of "America en la bibliografía Española (reseñas informativas)," *Historiografía y Bibliografía Americanistas* (Seville), 1 (1955)-33 (1988), #738.

Indexes, Cross-References, and Abbreviations

There are three indexes, all referring to item numbers rather than page numbers. The first index lists authors and editors, with titles included for important anonymous items (mostly collections). The second index lists persons and places treated in the entries; the third lists topics. Some overlapping is inevitable in the indexes, as when persons mentioned are also authors or editors; but since any clue to the location of something one is seeking might be helpful, the overlapping might be a virtue.

In the annotations there are many cross-references, all citing the item number of the entry referred to. The most prominent of these cross-references locate the individual essays, books, etc., listed in the entries for collections of various sorts (see, *e.g.*, #s 12 and 17), but many other cross-references tie together items—sometimes widely separated in the *Guide*—that treat the same subject.

Only the most frequently used journal titles have been abbreviated, and the abbreviations are designed to be recognizable as much as possible without the help of a list, which is nonetheless provided on p. xxix, below.

Selection

Several characteristics of the body of literature about Columbus make selectivity desirable and practicable. In the first place, many if not most of the articles and books on Columbus published prior to about 1875 must be excluded for obsolescence. Until the last third of the nineteenth century, most scholars did not have access to enough authentic documents to be knowledgeable in anything like the degree that was made possible by publications which began to appear in 1864 with the first volume of Pacheco's Collection (#4). There followed in 1875 the first printing of Las Casas' *Historia* (#59) and in 1892-96 the Italian government's *Raccolta* (#8), which drew together all of Columbus's available writings. In 1892 and 1902, the Duchess of Berwick published a number of additions to the canon (#s 6 and 9), chiefly letters by Columbus. Other important resources, such as the Fernández Duro edition of the *Pleitos* (#61) and the Assereto document (#188), swelled the sudden accumulation of material for study.

Thus, as a result of these newly available documents, Columbus scholarship in many important senses is only a century old. Perhaps in some senses it is only now beginning, to judge by the compendious recent scholarship in Spain and Italy (#s 15, 16, 18, 20, 22, 23, 24, 25), based largely on documents turned up in Genoa and in the still relatively under-explored Spanish archives.

There is another reason why radical selectivity is acceptable and advisable in this *Guide*. A large fraction of the publications on Columbus both in the nineteenth and the twentieth centuries has addressed moribund, dead, or irrelevant issues. For example, sensationalists never tire of pumping new life into the old issue of Columbus's nationality, though no responsible scholar in the past sixty years has doubted that he was Genoese. The documentary evidence for his Genoese origin, already mountainous in 1892, was set forth in definitive fashion in a worldwide, multi-language publication in 1931 (#197).

Among dead or moribund issues, the proposal that Columbus was Jewish has if anything been even more popular in the twentieth century than the nationality issue, although the very same body of documents that establishes him as Genoese also provides the only sensible answer to the question of his religious background. That answer is this: Columbus, like virtually any European of his time, might have had some Jewish forebears; but if he did, we do not know who they were. Some fascinating material has in fact turned up on the relationship between Columbus's writings and Jewish thought (see Juan Gil's article on the subject, #654). But this evidence must always be weighed against the ample and clear documentation that makes his forebears Ligurian Catholics as far back as his line can be traced.

The tendency for merely sensational or otherwise non-scholarly publications to gravitate toward emotional, ethnic, or religious issues has made it relatively easy to clear away irrelevant material and focus on publications inspired by a genuine desire to establish the truth about the great navigator. As an example of easily recognized irrelevance, now and then the nineteenth-century enthusiasm to prove Columbus a saint is revived in some form (see #s 655, 656, 657), although one wonders how anyone who has studied the subject in the slightest depth or with the slightest impartiality can possibly suppose—in the light of the mountainous evidence to the contrary—that Columbus was a saint in the sense of being a person radiant with holiness, unselfishness, or love of humanity.

And so it goes: although Columbus was indeed religious in his devotion to the institution of Christianity and in the sense of wishing to proselytize non-Christians, the bibliographer can set aside with few qualms most of what has been published on his saintliness. The same is true of most of what has been published about his putative Jewishness and his putative non-Genoese origin, because most writers on these subjects had made up their minds before they were ready to draw conclusions (the classic instance is Celso García de la Riega; see #s 189, 190, 191, 196). I have also deemed it wise to exclude most of the studies of the meaning of Columbus's curious signature, because the authors tend to engage in recondite mystical speculation that is not subject to critical evaluation. This leaves a somewhat more manageable body of material: editions of primary documents, biographies, and articles and monographs addressing genuine issues in the perennial quest for historical, geographical, nautical, political, social, and biological truth.

I have found it necessary, regrettably, to exclude literary works on Columbus, and studies of such works, because of the sheer volume of this material. Others, I trust, will be coming along with such bibliographies. The same is true of writings about Columbus celebrations and monuments. A short section on Columbus's portraits has been included (#s 677-85) because the discoverer's appearance is a significant aspect of his life, and it is possible that one or more of the surviving portraits transmit an authentic tradition of how he looked.

The Phases of Columbus Study

The study of the life and writings of Christopher Columbus divides conveniently into four phases. The first is the discovery of the relevant written documents produced during an initial period comprising Columbus's lifetime and its historical context. On these documents is based virtually all pertinent evidence except that provided by non-written records like archaeological findings, surviving portraits, etc. A second phase, beginning in 1823 with Spotorno's *Codice diplomatico* (#1), comprises the initial publication of the available primary documents, frequently in large collections. This phase was largely completed—barring a windfall of currently unknown documents—in the first decade of the twentieth century, although further documents have continued to appear occasionally throughout the century, most notably in the publications of Alicia Bache Gould (#260). Chapter I of this *Guide* records, among other collections, the major collections of primary documents, and Chapter II records specifically (1) publications containing individual texts of primary documents, edited and unedited, and (2) collections of edited texts of these documents.

The third phase, overlapping the second, consists of the *critical* editing of the primary documents on reliable principles. It began propitiously with the Italian government's great *Raccolta* of 1892-96 (#10), which includes the only fully annotated critical edition to date of the known writings of Columbus, by Cesare de Lollis (#46). Since 1896, of course, the canon of CC's writings has expanded through various discoveries of previously unknown documents (recorded in Chapter I, "Collections," and in Chapter II, "Texts"). Even more importantly, the principles of critical editing have been refined since the appearance of the 1892-96 *Raccolta*. Consequently a full re-editing is necessary even of the texts edited by de Lollis, and it is not clear that such a project is yet under way. Chapter III of this *Guide*, "Studies of Texts," consists largely of textual studies that in one way or another are preparatory for the achievement of sound critical editions.

The fourth phase, overlapping the three previous phases and still in progress like phases two and three, consists of attempts to define the details of Columbus's life and to solve the issues arising from these findings, *e.g.*, where and when he was born; what his religious background was; when and how and in what order he developed his "Enterprise of the Indies," *i.e.*, his project to sail west across the ocean to reach the Asiatic spice islands that the Portuguese had for years been trying to reach by sailing around Africa. Chapters IV, "Columbus's Life," and V, "Columbiana," record a selection of the published results of this fourth phase.

Since the third and fourth phases just discussed both depend on the thoroughness with which the tasks of previous phases have been carried out, Columbus historical scholarship will remain in a state of flux until the last significant surviving document has been discovered, properly edited, and analyzed, and the results subjected to whatever adjustments the findings of allied disciplines like archaeology require.

Editions

Much more vexing than the need to exclude many non-scholarly biographies, monographs, and articles from this *Guide* has been the problem of what to do with the large number of editions of primary documents, especially those of Columbus's *Letter* of March 4, 1493 announcing his discovery and of his *Journal of the First Voyage.* In many library catalogues the entries for these two items total half or more of all the entries on Columbus. The catalogue of the New York Public Library is a case in point.

In spite of this, the only critical edition of the canon of Columbus's writings with the textual notes indispensable to such an edition is the De Lollis edition, #46, cited above. At the time when that edition was made, as suggested above, the modern principles of copy-text and of textual emendation—the preconditions of a sound critical edition—had not evolved fully. This evolution was completed only in the 1950s, in a movement spearheaded by W. W. Greg's "The Rationale of Copy-Text," *Studies in Bibliography* (Charlottesville VA), 3 (1950): 19-36. For such an item as Columbus's *Letter* of March 4, 1493, which exists in more than one substantive primary document, there is no critical edition that both explains the choice of copy-text satisfactorily and provides a full account of the editor's choices among the variant substantive readings.

In this situation the selective bibliographer's choices among the various editions would be arbitrary under any circumstances, and my solution is simple: after the De Lollis *Raccolta* edition (#46), I have chosen the most complete modern edition, that of Consuelo Varela, whose *Cristóbal Colón: Textos y Documentos Completos* (#54)—though lacking in textual notes— represents the work of a careful textual scholar, as indicated by the editor's remarkably comprehensive introduction. Beyond this, I have included principally those editions which—in the front-matter, the notes, or the appendices—provide important new scholarly findings or important original scholarly evaluations of the work and its context. In addition, I have cited a French and a German translation of Columbus's writings (#s 42, 50) as well as various severely incomplete collections of these writings in English translation (*e.g.*, #s 1, 14, 29, 30), on the principle that a beginning scholar needs to start somewhere, and a faulty text is perhaps better than none to such a person. And of course I have included the edition of Columbus's *Journal of the First Voyage* (#57) that, with its scholarly apparatus, constitutes Volume 1 of the Italian Quincentennial Commission's magnificent *Nuova Raccolta Colombiana* (#27). This edition repeats Dr. Varela's transcription of the Las Casas manuscript already published in #54; here the identity of the copy-text is not a problem because the source is unique. It also publishes the De Lollis text of the *Letter* with considerable important apparatus, but still without establishing the principles of copy-text used and without word-by-word, line-by-line critical notes showing the reasons for the choices made between variant prospective readings. This makes it clear that in spite of the wealth of important material that the *Nuova Raccolta* is making available, its edition of Columbus's writings will not stand as a definitive critical edition.

Dr. Varela acknowledges this in part in her 1988 study "Observaciones para una edición de los diarios del primera y el tercer viaje colombinos," *Columbeis III* (Genoa: DARFICLET, 1988), pp. 7-17. Her conclusions in this article—which became available to me only after the text of this *Guide* had been printed—indicate that new considerations, introduced after the current editing projects were all under way, must guide the critical editor's choices among the marginal notes and the corrections that appear in the unique manuscript of Las Casas' transcription of the journals of Columbus's first and third voyages, in MS Vitrina 6-7 in the Biblioteca Nacional in Madrid. This attitude on the part of the most active and vigorous current editor of Columbus's writings suggests that we are still only on the threshold of an adequate critical edition of these writings.

Emphases

Beyond the basic matter of establishing the texts of the documents on which Columbus scholarship is based, the number of items to be found in the various sections of Chapter IV, "Columbus's Life," and Chapter V, "Columbiana," gives a rough—though only a rough—approximation of the scholarly interest shown over the years in the topics treated in these sections. It must be noted once more that the journalistic and emotional interest in Columbus's birthplace, nationality, possible Jewishness, and possible saintliness has generated an enormous number of publications on these topics, perhaps more than those devoted to all other topics in the field. Since most of these publications are in no sense scholarly, I have given these topics only enough attention to establish what scholars have learned about these matters through genuine interest in determining the truth. Beyond this I have let the notorious career of García de la Riega (#s 189, 190, 191, 196) serve to demonstrate how genuine scholarship is undercut by nationalistic or sentimental desires to establish unsupportable myths as fact.

The two topics to which the most scholarly attention has been given are the Landfall issue (#s 287-330) and the matter of Columbus's cosmology and cosmography (#s 487-586). Beyond these, interest has been scattered fairly evenly among the various other topics indicated in the section headings of the chapters. The absence of items on such topics as Columbus's Fourth Voyage (#s 364-67) and Bartholomew Columbus's life (#s 609, 613) suggests areas where further pertinent documents, and further study, can be hoped for. It must be remembered, however, that this is a selective bibliography, not an exhaustive one. Users seeking further treatments of a particular topic may well find what they want by consulting the indexes, notes, and bibliographies of the general treatments of Columbus's life, especially #s 132, 134, 137, 251, 252, and 441; and it would be hard to overstate the resources to be found in Vignaud's two compendia, #s 142 and 433.

A further caveat: the section titles of Chapters IV and V do not exhaust the important topics treated in the studies listed in this guide. An eminent example of this is the matter of Columbus's marginal notes in the books he read. The inferences to be made from these notes have developed into one of the most active current interests of Columbus study, as will be seen by examining the items listed under "postils" in the topic index. The other entries in the topic index will guide users to various matters which for one reason or another did not seem appropriate as section headings in Chapters IV and V.

Still, there are many riches in the items listed here that can only be found by personal examination of the items themselves, especially the books. The annotations of this *Guide* are a reasonably safe indication of the content of the articles, which normally must have a single, limited thesis in order to be published; but the safest course in establishing what may be found in the various books listed here is to consult the indexes of these books.

Implications for Future Columbus Scholarship

Clearly, the most crucial problem currently faced by Columbus scholars is the textual one: we need a fully annotated, fully authentic critical edition. It may be that Dr. Varela's texts are in many instances precisely what is needed, but this must be demonstrated by the publication of her textual notes and a full explanation of her choices among various prospective copy-texts and the various possible readings wherever more than one substantive reading is available. It may be that the texts which are being arranged for the late Fredi Chiappelli's *Repertorium Columbianum*, sponsored by the Center for Medieval and Renaissance Studies at UCLA, will constitute a modern critical edition as well as a complete set of facing-page English translations; if so, we have an additional important reason to look forward happily to the ultimate completion of this monumental project now under the leadership of Prof. Geoffrey Symcox.

It is my hope that this *Guide* has organized Columbus studies in a fashion that will aid future scholarship, initially by revealing what has already been accomplished, and then, perhaps, by revealing what has yet to be accomplished. To my mind the greatest current handicap to Columbus scholarship, besides the absence of a modern critical edition of

Columbus's writings, is the absence of essential journals and books even in major libraries. In the United States, library subscriptions to the most important Spanish-language journals—published in Spain and elsewhere—dwindled rapidly during the 1980s, not to mention the great difficulty of finding the more obscure journals, which frequently publish important articles. Also, a large proportion of the Genoese journals, monographs, and collections of studies, which together with the parallel Spanish publications dominate the field, are virtually impossible to find either in Spain or the United States. I hope that libraries will undertake to remedy the absence of Columbus journals and books, and that (to facilitate this), reprint houses will undertake to make the material listed in this *Guide* readily available to libraries on both sides of the Atlantic; only with some such program, in my opinion, can Columbus scholarship advance and prosper.

Sources

The sources of the information in this *Guide*, in addition to those mentioned above under "Annotations," are many. I must acknowledge the fundamental aid of Dr. Martin Torodash, who generously turned over to me the material assembled for his historiographical article, #769, and the similar material that he continued to assemble until 1982. Almost as important to me has been Dr. Simonetta Conti's compendious list of editions and studies in *Un secolo di bibliografia Colombiana 1880-1985*, #745.

In establishing my initial list, which included some 4000 items, I also derived much helpful information from the notes of Samuel Eliot Morison's *Admiral of the Ocean Sea* (#132) and *The European Discovery of America* (#251), as well as the *schede* and bibliographies of Paolo Emilio Taviani's *Cristoforo Colombo: la genesi della grande scoperta* (#441) and *I viaggi di Colombo: la grande scoperta* (#252). I consulted the bibliographies in Section VI of this *Guide*; the various historiographies in Section VII; and the indexes, reviews, summaries of scholarship, and notes in the *Hispanic American Historical Review* and the *American Historical Review*. I must also mention the scholarly notes in the other books and articles listed in the 780 entries of this *Guide*. Even after combing the various bibliographies, catalogues, and checklists described above, I still encountered many important items for the first time in these notes. A further source of titles for examination was the various library catalogues listed below.

To help in locating the items chosen for examination and possible annotation, I have had at hand copies of the Columbus listings in the catalogues of the following libraries and their collections: the Berio Civic Library in Genoa; the various libraries at Brown University, including especially the John Carter Brown and Rockefeller Libraries; the Widener and Pusey Libraries at Harvard; the Stirling and Mudd Libraries at Yale; the New York Public Library; the library of the Knights of Columbus in New Haven, Connecticut; the library of the Hispanic Society of America in New York; the libraries of SUNY Stony Brook and SUNY Albany; the libraries of the University of Michigan at Ann Arbor, including the William C. Clements Library and the Hatcher Graduate Library; the library of the University of Illinois at Urbana; the Duquesne University Library; the Hillman Library and Darlington Room Library at the University of Pittsburgh; the Pattee Library at Penn State; the Troy H. Middleton Library at Louisiana State University; the Suzzallo Library at the University of Washington, Seattle; and the libraries of the University of Florida at Gainesville and the University of Texas at Austin.

I have had the pleasure of visiting and using all these collections except the last two. I have also visited and used the collections of the Biblioteca Nacional in Madrid and the library of the University of Genoa. Beyond these, the vast and easily accessible resources of the Library of Congress, the National Union Catalogue, OCLC, and RLIN have furnished copious information and copies of numerous books and articles secured by interlibrary loan.

The errors in this *Guide* are my own.

Foster Provost
Duquesne University
and the John Carter Brown Library

Abbreviations

Abbreviations are self-explanatory in context, except (perhaps) "CC," used throughout the annotations to abbreviate "Christopher Columbus"; "LC," meaning the Library of Congress; "JCB," meaning the John Carter Brown Library; "*AOS*," meaning S.E. Morison's *Admiral of the Ocean Sea*, item #132 in this *Guide*; and "*Mug*," which labels an annotation adapted from item #735, Donald Mugridge's 1950 bibliography of American studies on Columbus. The numeral following "*Mug*" indicates the item number in Mugridge.

Where abbreviated, the full titles of Journals and Collections can be established from the following list.

Abbreviations of Titles of Journals and Collections

Many entries in this *Guide* include the full title of a journal or collection instead of an abbreviation; no attempt has been made to represent these journals and collections in this list. The numeral, *e.g.* #15, at the end of some entries in this list of abbreviations is the serial number in this *Guide* for the title abbreviated at the beginning of the entry.

Amer Hist Rev	*American Historical Review* (USA).
Anuario Estud Amer	*Anuario de Estudios Americanos* (Seville)
Atti I Conv Internaz Stud Col	*Atti del Convegno Internazionale di Studi Colombiani, 1973* (Genoa, 1974), #15.
Atti II Conv Internaz Stud Col	*Atti del II Convegno Internazionale di Studi Colombiani, 1975* (Genoa, 1977), #16.
Atti III Conv Internaz Stud Col	*Atti del III Convegno Internazionale di Studi Colombiani, 1977* (Genoa, 1979), #18.

Atti Soc Ligur Stor Pat	*Atti della Società Liguriana di Storia Patria* (Genoa).
Bol Civico Ist Col	*Bollettino del Civico Istituto Colombiano* (Genoa).
Bol R Acad Hist	*Boletin de la Real Academia de la Historia* (Madrid).
Bol Soc Geog Lisboa	*Boletim da Sociedade Geografica de Lisboa* (Lisbon).
Boll R Soc Geog Ital	*Bollettino Reale Società Geografica Italiana* (Rome).
Boll Soc Geog Ital	*Bollettino Società Geografica Italiana* (Rome).
Bull Am Meteorol Soc	*Bulletin of the American Meteorological Society* (Milton MA, USA).
Bull N Y Pub Lib	*Bulletin of the New York Public Library* (USA).
Canadian Hist Rev	*Canadian Historical Review* (Toronto, Canada).
Cath Hist Rev	*Catholic Historical Review* (Wash. DC, USA).
Columbus and his World	*Columbus and his World: Proceedings, First San Salvador Conference, held at San Salvador Island, Bahamas, Oct 30 – Nov. 3, 1986* (Ft. Lauderdale FL, USA, 1987), #26.
Geog Jour	*Geographical Journal* (London).
Geog Rev	*Geographical Review* (New York, USA).
Giorn Storico e Lett Ligur	*Giornale Storico e Letterario della Liguria* (Genoa).

Hisp Amer Hist Rev	*Hispanic American Historical Review* (USA).
Hist y Bibl Amer	*Historigrafía y Bibliografía Americanistas* (Seville).
Jour Amer Geog Soc	*Journal of the American Geographical Society* (NY, USA)
Papers Bib Soc Amer	*Papers of the Bibliographical Society of America* (USA).
Pres Ital Andalu I	*Presencía Italiana en Andalucía, Siglos XIV-XVII. Actas del I Coloquio Hispano-Italiano, 1983.* (Seville, 1985), #20.
Pres Ital Andalu II	*La presenza italiana in Andalusia nel basso medioevo. Atti del II Colloquio Italiano-Spagnolo, Roma, 25-27 maggio 1984* (Bologna, 1986), #22.
Proc Amer Antiq Soc	*Proceedings of the American Antiquarian Society* (Worcester MA, USA).
Proc Int Cong Amer	*Proceedings of the International Congress of Americanists* (various countries).
Proc Mass Hist Soc	*Proceedings of the Massachusetts Historical Society* (Boston MA, USA).
Raccolta	*Raccolta di documenti e studi pubblicati dalla R. Commissione colombiana pel quarto centenario della scoperta dell'America* (Rome).
Rev Gen Marina	*Revista General de Marina* (Madrid).
Rev Geog	*Revue de Geographie* (Paris).
Rev Geog Amer	*Revista Geográfica Americana* (Buenos Aires).

Rev Hist Amer	*Revista de Historia de América* (Mexico City).
Rev Indias	*Revista de Indias* (Madrid).
Riv Geog	*Rivista Geografica Italiana* (Florence).
Studi Colombiani	*Studi Colombiani: Atti del Convegno Internazionale di Studi Colombiani, 1951* (Genoa, 1952), #12.
Temi Colombiani	*Temi Colombiani, Scritti in Onore di Paolo Emilio Taviani.* (Genoa, 1986), #24.
Terr Incog	*Terrae Incognitae* (USA).
Trans Am Phil Soc	*Transactions of the American Philosophical Society* (Philadelphia, USA).

COLUMBUS

An Annotated Guide to the Scholarship

on His Life and Writings, 1750 to 1988

I

Collections of Sources, Texts, and Studies

[1] **Giovanni Batista Spotorno.** *Codice diplomatico Colombo-Americano ossia Raccolta di documenti originali e inediti, spettanti a Cristoforo Colombo alla scoperta ed al governo dell'America.* Genoa: Ponthenier, 1823. 348 pp.

First major compendium of CC documents, anticipating Navarrete (#2) and the *Raccolta* of 1892-96 (#8). After an introductory biographical memoir, pp. i-xxx, prints the documents in Spanish or Latin, with facing-page Italian translations, including (a) letters to, from, or in behalf of CC; (b) privileges granted to CC; (c) royal cedulas relative to CC; and (d) the papal bull of Alexander VI (4 May 1493) fixing the line between Spanish and Portuguese areas of conquest.

English ed. of the translations: *Memorials of Columbus; or a Collection of Authentic Documents of that Celebrated Navigator,* etc. London: Treuttel and Winter, 1823. 256 pp.

[2] **Martín Fernández de Navarrete,** ed. *Colección de los viajes y descubrimientos que hicieron por mar los Españoles desde fines del siglo XV,* etc. Madrid: Imprenta Real, 1825. 2 vols.

Vol. 1: *Viajes de Colón, Almirantazgo de Castilla.* Contents: Pp. i-cli, Historical Introduction. Pp. 1-329, narratives, letters, and other documents concerning CC's four voyages. Voyage 1: (a) pp. 1-166, the Las Casas abstract of CC's *Journal* of the First Voyage, pub. here for the first time; (b) CC's *Letter to Santangel* of 4 Mar. 1493; (c) Latin translation of CC's *Letter to Rafael Sánchez.* Voyage 2: (a): the *Letter* of Dr. Chanca to Seville; (b) CC's memorandum to the monarchs from Isabela, 30 Jan 1494, sent by Antonio de Torres. Voyage 3: (a) CC's narrative of the 3rd voyage sent to the monarchs from Hispaniola; CC's letter to Prince Juan's nurse (1500). Voyage 4: (a) two letters from the monarchs to CC concerning the voyage, and one to the captains of the Portuguese armada; (b) CC's brief narrative of the voyage to the monarchs: account of the route along Central America, of the gold found, of the crews, and of the loss of the ships; (c) CC's Letter to the monarchs from Jamaica of 7 Jul 1503, and Diego Méndez's acct. from his will. Pp. 330-52, 4 letters from CC to Fr. Gorricio and 11 to his son Diego. After p. 352, a map of the 4 voyages. Pp. 353-429, Appendix of documents relative to the prerogatives and jurisdiction of the Major Admiralcy of Castile. Pp. 430-455, summary and index of the volume. At the end of each document, N. tells its location and the date when he saw and copied it.

Vol. 2: *Documentos de Colón y de las Primeras Poblaciones.* Contents: Pp. 1-371, 172 docs. relating to CC's career, from Rey San Fernando's letter of 1251 a.d. opening trade with Genoese merchants to the notices authorizing transfer of CC's body from Santo Domingo to Havana in 1795-6. Pp. 373-438, Appendix of 21 docs cited in the Intro. to Vol 1. Pp. 439-55, a chronological index of these documents. Unlike Vol. 1, Vol. 2 does not cite the location of docs or of the date seen and copied.

[3] **Martín Fernández de Navarrete, Miguel Salvá, and Pedro Sainz de Baranda.** *Colección de documentos inéditos para la historia de España.* Madrid: Calero, 1842-95. 112 vols.

Collection begins with documents related to Hernan Cortés, and includes material concerning Diego C. and his administration.

[4] **Joaquín F. Pacheco, Francisco de Cárdenas, and Luis Torres de Mendoza.** *Colección de documentos inéditos relativos al descubrimiento, conquista y colonización de las posesiones españoles en América y Oceania, sacados en su mayor parte del Real Archivo de Indias.* Series 1, 42 vols. Madrid: de Quivos, 1864-84. Series 2, 25 vols. Madrid: de Quivos, 1885-1932.

These volumes record the texts of 3807 previously unpublished documents, mostly from the Archive of the Indies in Seville, relating to CC's discovery and the consequent colonial developments. Includes 530 docs from the years 1474-1506 directly relating to CC's career and to consequent voyages and governmental actions. A prime resource for the study of CC's life.

To use these volumes efficiently the Index is indispensable: Ernesto Schäffer, *Indice de la colección de documentos inéditos de Indias* (Madrid: Instituto "Gonzalo Fernández de Oviedo," 1946), 2 vols. Vol. 1 provides an alphabetical index of persons mentioned in both series; Vol. 2, a chronological list of documents in both series. 3807 docs, 1251 – 1883 a.d.

[5] **María del Rosario Falcó y Osorio, Duquesa de Berwick y Alba,** ed. *Documentos escogidos del archivo de la Casa de Alba.* Madrid: Manuel Tello, 1891.

See pp. 201-228.

[6] **María del Rosario Falcó y Osorio, Duquesa de Berwick y Alba,** ed. *Autógrafos de Cristóbal Colón y papeles de América.* Madrid: Sucesores de Rivadeneyra, 1892. 203 pp.

Prints a wealth of previously unpublished primary documents found in the family collection, including 16 to, by, or about CC, 5 to, by, or about Diego his son, and many others. Index of proper names.

[7] **Joaquín Torres Asensio,** ed. & trans. *Fuentes Históricos sobre Colón y América.* Madrid: de Sales, 1892. 4 vols.

Vol. 1, pp. 17-96, Spanish translations of 43 letters by Peter Martyr touching on CC; pp. 97-380, Spanish translation of PM's *First Decade,* which contains the historical material on CC.

[8] **Italy. R. Commissione Colombiana.** *Raccolta di documenti e studi pubblicati dalla R. Commissione colombiana pel quarto centenario della scoperta dell'America.* Rome: Ministro della Pubblica Istruzione, 1892-96.

Part I (1892-94). Cesare de Lollis, ed. *I Scritti di Cristoforo Colombo.* To date, the only critical edition of all the known writings with full apparatus. 3 vols., in 4 tomes. For contents, see individual entry, #46.

Part II, vol. 1 (1896). L. T. Belgrano and M. Staglieno, eds. *Documenti Relativi a Cristoforo Colombo e alla sua Famiglia.* Chapters, with documents: (1) Giovanni Colombo and his son Domenico; (2) Domenico C., CC's father; (3) Domenico C., 1470-72: Primary documents naming Domenico's son Cristoforo; (4) Notices of D.C. and his family, 14 Oct 1470 – 1474, especially in Savona; (5) More on CC and part of his family in Savona; (6) D.C. back in Genoa; (7) CC, Admiral of Spain, and documents regarding him personally; (8) Diego, son of CC and Felipa Moniz, and their testaments; (9) Bartholomew C. and his brother Diego; (10) Fernando, son of CC and Beatriz Enríquez; (11) María de Toledo, widow of Diego, and their descendants. Columbus family tree, p. 281; Index, p. 293.

Part II, vol. 2 (1894). L. T. Belgrano and M. Staglieno, eds. *Il Codice dei Privilegi di Cristoforo Colombo, Edito Secondo i Manoscritti di Genova, di Parigi, e di Providence.* This item separately entered, #32.

Part II, vol. 3 (1894). 3 monographs, entered separately in this guide: (1) "Quistioni [sic] Colombiane," by Cornelio Desimoni, #760; (2) "Cristoforo Colombo e i Corsari Colombo,"; by Alberto Salvagnini, #186; (3) "I Ritratti di Cristoforo Colombo," by Achille Neri, #682. Fourth monograph, not entered in this Guide, "Le Medaglie di Colombo," by Umberto Rossi.

Part III. Guglielmo Berchet, ed. *Fonti Italiane per la Storia della Scoperta del Nuovo Mondo.*

Vol. 1 (1892). *Carteggi Diplomatici* (Diplomatic Correspondence). Critical edition of correspondence concerning CC and the New World, from various Italian cities, comprising 16 items from Rome, 1493-1530; 48 from Venice, 1497-1536; 18 from Ferrara, 1493-1530; 24 from Mantua, 1494-1534; 5 from Milan, 1493-1497; 8 from Genoa, 1518-1535; 7 from Florence, 1513-1525.

Part III, vol. 2 (1893). Guglielmo Berchet, ed. *Narrazioni Sincrone* (contemporary accounts of CC and/or the newly found lands). Reproduces 194 fully annotated narrative accounts, 1493-1555, including those of Tribaldo di Rossi, Giuliano Dati, Peter Martyr, Battista Fregoso, Niccoló Scillacio, Michele da Cuneo, Marcantonio Sabellico, Amerigo Vespucci, Antonio Gallo, Agostino Giustiniani, Alessandro Geraldini, Giralomo Fracastoro, and more than 100 other narrators.

Part IV, vol. 1 (1893). E. A. d'Albertis. *Le Costruzioni Navali e l'arte della Navigazione al Tempo di Cristoforo Colombo.* 240 pp. Entered separately in this guide, #697.

Part IV, vol. 2 (1892). 2 monographs, entered separately in this Guide: (1) Timoteo Bertelli, "La Declinazione Magnetica e la sua Variazione nello Spazio Scoperte da Cristoforo Colombo," pp. 7-99 (#572); (2) Vittore Bellio, "Notizia delle Più Antiche Carte Geografiche che si Trovano in Italia Riguardanti l'America," pp. 101-221, #487.

Part V, vol. 1 (1894). Gustavo Uzielli. *La Vita e Tempi di Paolo del Pozzo Toscanelli. Ricerche e Studi.* 745 pp. Entered separately in this Guide, #528.

Part V, vol. 2 (1894). (1) Monograph by Giuseppe Pennesi, "Pietro Martire d'Anghiera e le sue Relazioni sulle Scoperte Oceaniche," pp. 5-109; (2) Luigi Hugues, "Amerigo Vespucci," summary notice, pp. 111-150, entered separately in this Guide, #720; (3) Vincenzo Bellemo, "Giovanni Caboto," critical notes, pp. 151-218; (4) Luigi Hugues, "Giovanni Verrazano," summary notice, pp. 219-51; (5) Luigi Hugues, "Juan Bautista Genovese," summary notice, pp. 253-62; (6) Prospero Peragallo, "Sussidi Documentari per una Monografia su Leone Pancaldo," pp. 263-306.

Part V, vol. 3 (1894). 2 monographs: (1) Andrea da Mosto, "Antonio Pigafetta e le sue Regole sull'Arte del Navigare," pp. 7-131; (2) Marco Allegri, "Di Gerolamo Benzoni e della sua *Historia del Mondo Nuovo*," pp. 133-54.

Part VI (1893). Giuseppe Fumagalli and Pietro Amat di S. Filippo. *Bibliografia degli Scritti Italiani o Stampati in Italia sopra Cristoforo Colombo, la Scoperta del Nuovo Mondo, e i Viaggi degli Italiani in America.* 1400 Items. 217 pp. Entered separately in this guide, #732.

[9] **María del Rosario Falcó y Osorio, Duquesa de Berwick y Alba,** ed. *Nuevos autógrafos de Cristóbal Colón y relaciones de ultramar.* Madrid: Berwick, 1902. 204 pp. (This work supplements #6.)

This second collection of hitherto unpublished docs, mostly accounts of early voyages and travels in the new Spanish lands, begins with 10 letters, memoranda, etc., directly concerned with CC's personal or public life, including a letter from CC to his son Diego on 29 April of an unnamed year, and 7 to Fr. Gorricio, 1498-1501. Plates, index of proper names.

[10] **Cecil Jane.** *Select Documents Illustrating the Four Voyages of Columbus including those Contained in R. H. Major's "Select Letters of Christopher Columbus."* London: Hakluyt Society, 1930-33. 2 Vols.

Vol. 1, *The First and Second Voyages.* 1930. Hakluyt Society pubs., series 2, no. 65. Documents, with facing-page English translation. Voyage 1: CC's *Letter to Santangel* of 4 Mar 1493. Voyage 2: Dr. Chanca's letter to the city of Seville; CC's memorandum to the monarchs, sent by Antonio de Torres; and chaps. 123-31 of Andrés Bernáldez's *History of the Catholic Sovereigns.*

Jane's elaborate introduction, "The Objective of Columbus," pp. xiii-cxxii, takes a skeptical position toward the opinions that CC was in any degree learned, that he was capable of profound intellection of any kind, or that he was even capable of writing the notes attributed to him in *Imago Mundi*, etc. Jane acknowledges CC's intent to reach Cathay, but strongly suggests that his confidence of success derived from a quasi-fanatical religious conviction of his own status as God's chosen instrument. Pp. cxxiii-cl, Textual Introduction.

Vol. 2, *The Third and Fourth Voyages.* 1933. Hakluyt Society pubs., series 2, no. 70. Edition completed by E. G. R. Taylor after Jane's death. Jane's Introduction, "The Negotiations with Ferdinand and Isabella," pp. xiii-lxxv, attacks the myth that the delay in accepting CC's Enterprise reflected perversity in the court of Castile, and attributes it instead to the Moorish war and CC's high demands. See also E.G.R. Taylor's essay, "Columbus and the World Map," pp.

lxxvi-lxxxiv, which amplifies Jane's attitude in the intro to v. 1. She begins with the premise that CC "was a self-educated, emotional, unpopular man, prone to self-pity, who clung tenaciously and fervently to a quite mistaken cosmographical theory." She proceeds to explore the medieval ideas that led CC to this mistaken cosmography.

Textual introduction to docs. pp. lxxxv – lxxxix. 4 docs, with facing page English translations: (1) Voyage 3: Letter of CC to their Majesties; (2) Voyage 3: Letter of CC to the Nurse; (3) Voyage 4: Letter of CC to their Majesties; (4) Voyage 4: Letter of Diego Méndez.

[11] **Instituto Hispano-Cubano de Historia de América.** *Documentos Americanos del archivo de protocoles de Sevilla, siglo XVI.* Madrid: Typografía de Archivos, 1935. 518 pp.

An annotated list of the 16th-c documents that have direct bearing on the study of American history. Classifies the documents chronologically for each place of issue. Indexes of subjects, persons, and places, with variants of names.

Rina Ferrarelli Provost

[12] **Genoa. Comitato Cittadino per le Celebrazioni Colombiane.** *Studi Colombiani: Atti del Convegno Internazionale di Studi Colombiani, 1951.* Genoa: Stabilimento Arti Grafiche ed Affini, 1952. 3 vols.

Vol. 1: Account of the meeting, including the lectures. Vols. 2 & 3: essays based on the papers presented. Essays in Vol. 2 grouped under themes: CC's culture; CC and the ambience of the Castilian court; the evolution of exploration and knowledge of America; and CC and the Indies. Vol. 3 contains essays on various themes, and 21 appendices treating various further matters of interest.

The following essays are entered in this Guide: P. Revelli, "L'Italianità di CC," 2: 9-38, #642; S.E. Morison, "C. as a Navigator," 2: 39-48, #668; H. Winter, "Bemerkungen zur Navigation von Kolumbus und der seiner Zeit," 2: 49-54, #669; E. Rodríguez Demorizi, "Colón y el refranero," 2: 55-7, #643; A. Davies, "Origins of Colombian Cosmology," 2: 59-67, #508; G. Raffo, "Sulle postille di C. relative alla storia Romana," 2: 69-75, #641; E. Jos, " *El Diario de Colón:* su fondamental autenticidad," 2: 77-79, #98; T.C. Giannini, "Psicologia Colombiana: base psicologica del disegno di C." 2: 113-118, #478; P. Scotti, "Concetti etnologici di C." 2: 119-25, #155; A. Muro Orejón, "Un autógrafo de CC," 2: 127-35, # 48; A. Tonneau, "L'enigme des chiffres de CC," 2: 137-80, #460.

J. Colomer Montset & P. Catalá Roca, "Las escrituras de CC y consideraciones sobre sus firmas," 2: 181-205, #458; P. Leturia,

"Ideales político-religiosos de C. en su carta institucional del 'Mayorazgo': 1498," 2: 249-75, #153; A. Blum, "Un but des voyages de CC," 2: 211-16, #438; P. Catalá Roca, "Sobre los italianismos observados en la carta de C. a Santangel," 2: 283-90, #459; J. Guillen Tato, "La parla marinera en el Diario del primer viaje de CC," 2: 291-93, #457; A. Palomeque Torres, "Ambiente político y científico que rodeo al futuro Almirante ... en la España de los Reyes Católicos," 2: 303-355, #228; A. Angulo Pérez, "La odisea colombina y el destino de una empresa," 2: 357-61, #382; M. Freixa Ubach, "El pasaporte y la carta de presentación que los Reyes Católicos entregaron a CC en 1492," 2: 363-67, #68; P. Catalá Roca, "Les monjes que acompanaron a C. en el 2o viaje," 2: 371-90, #340; J. Colomer Montset, "Las capitulaciones de Santa Fé registradas en el archivo de la corona de Aragón, en Barcelona," 2: 391-403, #41.

C. Verlinden, "C. e les influences mediévales dans la colonisation de Amérique," 2: 407-418, #383; I.O. Bignardelli, "Circa la necessità d'una nuova rivalutazione delle fonti della storia colombiana e della scoperta," 2: 443-49, #766; E. Lunardi, "L'importanza del Monastero di S. María de la Rábida nella genesi della scoperta dell'America," 2: 451-67, #439; A. Codazzi, "Di una versione italiana manoscritta della lettera di CC al 'Thesorero de sua Mta,'" 2: 469-78, #97; M. Destombes, "Une carte interessant les études colombiennes conservée à Modene," 2: 479-89, #556; G. Praticò, "La scoperta dell'America nei documenti dell'archivio de stato di Mantova," 2: 489-90, #84; A. Mori, "Giuseppe Pagni e il suo manoscritto inedito su CC e Amerigo Vespucci," 2: 491-4, #116; G. Caraci, "I problemi vespucciani e i loro recenti studiosi," 2: 495-552, # 723.

E. Rossi, "Scritti turchi su CC e la scoperta dell'America," 2: 563-66, #767; G. Imbrighi, "Sui rapporti tra le 'Historie' attribuite a D. Fernando Colombo e la 'Historia' scritta del Vescovo Bartolomeo de las Casas," 2: 567-79, #115; P. Revelli, "Nuovo contributo di R. Levillier allo studio delle fonti più antiche sui viaggi di Vespucci,: 2: 649-71, #724; F. Domínguez Compañy, "C. y los indios," 2: 691-96, #152.

A. Alvarez Pedroso, "Los restos mortales del descubridor de América," 3: 15-23, #691, "El verdadero retrato de CC," 3: 25-29, #684, and "Recuerdos colombinos de la Republicana Dominicana," 3: 31-44, #151; G. R. Crone, "Fra Mauro's Representation of the Indian Ocean and the Eastern Islands," 3: 57-64, #551; A. Brian, "La fauna marina ai tempi di CC," 3: 71-74, #589; G. Pesce, "I medici di bordo ai tempi di CC," 3: 75-83, #590; A. Benedicenti, "CC e la medicina," 3: 117-24, #588; E. Remotti, "Concetti antropologici di CC," 3: 243-47, #268; G.E. Broche, "CC a-t-il atteint l'Islande?" 3:

249-56, #215; G. Odoardi, "Il processo di beatificazione di CC," 3: 261-72, #657.

Index, pp. 576-7.

[13] **Academia de Geografía e Historia de Costa Rica.** *Colección de documentos para la historia de Costa Rica relativos al cuarto y último viaje de Cristóbal Colón,* with Prologue and Introduction by Jorge A. Lines. San José, Costa Rica: Atenea, 1952. 331 pp.

79 documents, under three headings: (1) 53 royal cedulas, albalás, and other docs by CC and others, relating to the 4th voyage; (2) accts of 4th voyage by Peter Martyr, Ferdinand Columbus, Bernáldez, Gómara, Oviedo, Las Casas, and Herrera; (3) 9 principal commentaries on the locale and identity of the section of Central America that CC designated as "Cariay."

[14] **Samuel Eliot Morison,** trans. & ed. *Journals and Other Documents on the Life and Voyages of Christopher Columbus.* New York: Heritage, 1963. 417 pp.

Part I: Docs on CC's early life, incl. Assereto; Toscanelli corresp.; "preparations" for Voyage 1; Capitulations and related decrees. Part II, Voyage 1: *Journal.* Part III, Voyage 2: nar. by CC, de Cuneo, Syllacio, Ferdinand Columbus. Part IV, Voyage 3: Journal, CC's letters, royal mandate of restitution. Part IV, Voyage 4: Royal instructions; roster and payroll; acct by Ferdinand Columbus; CC's "Lettera Rarissima"; will of Diego Méndez. Maps, Appendix, Index.

[15] *Atti del Convegno Internazionale di Studi Colombiani, 1973,* ed. Renzo Lagomarsino. Genoa: Civico Istituto Colombiano, 1974. 175 pp.

Essays deriving from the meeting in Genoa, 13 & 14 Oct. 1973. Prolusion, P. E. Taviani, pp. 19-33. Essays included in this Guide by Joaquín Arce, "Problemi linguisitic inerenti nel Diario de CC," #462, and Osvaldo Baldacci, "Le cartonautica medioevale precolombiana," #513.

[16] *Atti del II Convegno Internazionale di Studi Colombiani, 1975.* Genoa: CIC, 1977. 438 pp.

Essays deriving from the meeting in Genoa, 6 & 7 Oct., 1975. Essays in this Guide (short titles) by Elio Migliorini, "Gli studi colombiani," #519; Aldo Agosto, "Due nuovi documenti," #43; Raquel Soeiro de Brito, "Les îles de l'Atlantique," #423; Caterina Barlettaro and Ofelia Garbarino, "Il fondo cartografico," #518; Ilaria Luzzana Caraci, "La postilla colombiana B858," #164; Ernesto Lunardi, "Le 'Capitulaciones de Santa Fé,'" #231.

[17] **Fredi Chiappelli,** ed. *First Images of America*. Berkeley and Los Angeles: Univ. of California Press, 1976. 2 vols.

56 essays recording an international congress at UCLA in 1975 devoted to the initial impact of the New World on the Old, under these headings: 1. Introduction: Renaissance and Discovery; 2. Angles of Perception: Myth and Literature; 3. The Politics of Conflict; 4. Governing the New World: Moral, Legal, and Theological Aspects; 5. Images in the Arts; 6. Books; 7. Language; 8. The New Geography; 9. The Movement of People; 10. Science and Trade. 10. Epilogue.

Selected essays recorded in this Guide, T. Adams, "Discovery and the Invention of Printing," #419; M. Battlori, "Papal Division of the World," #584; R.L.Benson, "Medieval Canonistic Origins," #420; A. Gerbi, "Earliest Accounts on the New World," #592; T. Goldstein, "Impulses of Italian Renaissance Culture," #421; S.J. Greenblatt, "Learning to Curse," #273; E.J. Hamilton, "What the New World Gave ... the Old," #446; R. Hirsch, "Printed Reports on Early Discoveries," #274; H.B. Johnson, "New Geographical Horizons, Concepts," #514; U. Lamb, "Cosmographers of Seville," #515; C. E. Nowell, "Old World Origins," #422; D.B. Quinn, "New Geographical Horizons, Literature," #539; F. Rogers, "Celestial Navigation," #516; J. Snyder, "Jan Mostaert's West Indian Landscape," #275; N. Thrower, "New Geographical Horizons: Maps," #517; L. Weckmann, "The Alexandrine Bulls," #585.

[18] *Atti del III Convegno Internazionale di Studi Colombiani, 1977*. Genoa: CIC, 1979. 734 pp.

Essays deriving from the meeting in Genoa, 7 & 8 Oct., 1977. Entered in this Guide (short titles), A. Agosto, "La scoperta nel 1610 della lettera di Colombo ... San Giorgio," #53; G. Balbis, "Per la storia dei Colombo in Liguria," #610; O. Baldacci, "Tecnica nautica," #674; Laura Balletto, "Chio nel tempo di Cristoforo Colombo," #202; L. Bernardis, "Le bolle Alessandrine," #165; A. Boscolo, "Ricerche su Cristoforo Colombo," #771; Francesca Cantù, "Cristoforo Colombo ... di las Casas," #166; F. Castellano, "Domestici di Cristoforo e Diego Colombo," #628; G. DePaoli and M. G. Lucia, "Saggio di bibliografia," #772; G. Galliano, "Guida alla opere di interesse Colombiano conservate presso le pricipali biblioteche di Genova," #742; G.O. Galfrascoli, "Náutica y ciencias geográficas," #485; M.V.de Gerulewicz, "L'America agli occhi dei scopritori," #449; F. Morais do Rosario, "A escala de Colombo em Lisboa," #278.

[19] **John Parker and Louis De Vorsey,** eds. *In the Wake of Columbus: Islands and Controversy.* Detroit: Wayne State Univ. Press, 1985. 272 pp.

Reprinted from *Terrae Incognitae,* 15 (1985). Articles and texts dealing with the Landfall problem. Authors and short titles, entered in the appropriate places in this Guide: John Parker, "The Columbus Landfall Problem," #315; Pieter Verhoog, "Columbus Landed on Caicos," #320; Oliver Dunn, "Columbus's First Landing Place," #316; Robert H. Fuson, "The Diario de Colón," #103; James E. Kelley, Jr., "In the Wake of Columbus," #317; Arne B. Molander, "A New Approach," #318; Robert H. Power, "The Discovery of Columbus's Island Passage," #319; Oliver Dunn, "The Diario," #55.

[20] **Bibiano Torres Ramírez and José Hernández Palomo,** eds. *Presencía Italiana en Andalucía, Siglos XIV-XVII. Actas del I Coloquio Hispano-Italiano.* Seville: Escuela de Estudios Hispano-Americanos, 1985. 277pp.

A collection of essays, based on presentations at the conference, which provide detailed background for a study of the ambience in which CC moved while he was in Andalusia during his years in Spain. Included in this guide, three essays treating CC specifically: P.E. Taviani, "Si perfezionò in Castiglia il grande disegno di C.," pp. 1-19, #233, and "Ancora sulle vicende di C. in Castiglia, pp. 221-48, #234; and A. Boscolo, "Il genovese Francesco Pinelli," pp. 249-65, #172.

[21] **Alberto Boscolo.** *Saggi su Cristoforo Colombo.* Rome: Bulzoni, 1986. 108 pp.

Essays entered in this Guide: "Il genovese Francesco Pinelli amico a Seviglia di CC," #172; "Gli Esbarroya amici a Córdova di CC," #236; "Diego e Fernando Colombo paggi alla corte dei Re Cattolici," #630; "Fiorentini in Andalusia all'epoca di CC," #175; and "CC, La Isabela, e il memoriale Torres," #174.

[22] **Alberto Boscolo and Bibiano Torres,** eds. *La presenza italiana in Andalusia nel basso medioevo. Atti del II Colloquio Italiano-Spagnolo, Roma, 25-27 maggio 1984.* Bologna: Cappelli, 1986. 242 pp.

Includes the following essays entered in this guide: P.E. Taviani, "Brevi cenni sulla residenza di C. in Andalusia," pp. 7-12, #237; A. Boscolo, "Gli Esbarroya amici a Cordova di CC," pp. 13-19, #236; A. Albonico, "Bartolomeo C., adelantado mayor de las Indias," pp. 51-70, #613; C. Varela, "El intorno florentino de CC," pp.125-34, #179; G. Ferro, "I luoghi di C. e della sua familia in Liguria," pp. 135-42, #614; I. Luzzana Caraci, "Punti di contatto e divergenze tra la storiografia vespucciana e quella colombiana," pp. 143-55, #729; O. Baldacci, "La geocarta come documento storico colombiano," pp. 157-68, #522; and J.E. López de Coca Castener, "Publicidad en torno al tercer viaje colombino," pp. 233-42, #363.

[23] *Columbeis I.* Genoa: Istituto di Filologia Classica e Medievale, 1986. 222 pp.

Includes the following essays entered in this Guide: Simonetta Conti, "Orientamenti bibliografici colombiani," #777; Jacques Heers, "Le projet de Christophe Colomb," #779; Gabriella Moretti, "Nec sit terris ultima Thule (la profezia di Seneca sulla scoperta del Nuovo Mondo)," #429; Paola Navone, "Colombo e il 'Bestiario' dell'oriente meraviglioso," #466; Stefano Pittaluga, "Il 'vocabolario' usato da CC," #467; and Rosanna Rocca, "CC e la 'Isla de Córcega,'" #524.

[24] **Università degli Studi di Genova.** *Temi Colombiani, Scritti in Onore di Paolo Emilio Taviani.* Genoa: Edizioni Culturali Internazionali, 1986. 376 pp.

Tome 3 of *Scritti Colombiani.* Essays mostly related to CC, constituting year 13 of the Annals of the Faculty of Political Science. Authors and short titles of constituent essays included in this Guide: O. Baldacci, "Il segreto di C," #430; A. Boscolo, "Diego e Fernando C," #630; M. Conti, "Le postille de CC," #369; G. Ferro, "Storia delle esplorazione e geografia," #778; J.M. Martínez Hidalgo, "Las naves de los cuatro viajes," #708; G.M. Ugolini, "Il primo viaggio: costo dell'impresa," #281; A. Unali, "L'oro nei primi due viaggi," #253; C. Varela, "John Day, los genoveses, y Colón," #180.

[25] *Columbeis II.* Genoa: DARFICLET, 1987. 418 pp.

Includes the following essays entered in this Guide: Aldo Agosto, "In quale 'Pavia' studiò Colombo?" #206; Giuseppe Bellini, " . . . 'Andavan todos desnudos . . . ': alle origini dell'incontro tra l'Europa e l'America," #182; Giorgio Bertoni, "Appunti sugli italianismi linguistici de C.," #468, and "L'occhio, l'ancora, la scrittura, lo sguardo dell'almirante," #469; Mario Damonte, "Le lingue de CC," #470; Anna M. Mignone, "Index verborum Columbianus: Il *Diario di Bordo,*" #471; Giovanna Petti Balbi, "La scuola a Genova e CC," #207; Stefano Pittalugo, "CC Amanuense (e il suo incunabolo del 'Catholicon' di Giovanni Balbi)," #472; Rosanna Rocca, "Il lessico di Michele da Cuneo," #473.

[26] **Donald T. Gerace,** ed. *Columbus and his World: Proceedings, First San Salvador Conference* (held at San Salvador Island, Bahamas, Oct 30 – Nov. 3, 1986). Ft. Lauderdale, FL: College Center of the Finger Lakes, 1987. 359 pp.

Essays, based on the presentations at the conference, entered in the appropriate places in this Guide (short titles): by P.E. Taviani, "Why We are Favorable for Watlings," #330; Consuelo Varela, "Florentines' Friendship and Kinship with CC," #181; Delno C. West, "Scholarly Encounters with C's Libro de las Profecías," #107; Foster Provost, "C's Seven Years in Spain," #238; Carla Rahn Phillips, "Spanish Ships in the Age of Discovery," #709; Gaetano Ferro, "C. and his Sailings," #283; Georges A. Charlier, "Value of the Mile," #525; James E. Kelley, Jr., "The Navigation of C.," #675; Arne B. Molander, "Egg Island is the Landfall," #327; Robert H. Fuson, "The Turks and Caicos Islands as Possible Landfall Sites," #324; Mauricio Obregón, "C's First Landfall: San Salvador," #328; Donald T. Gerace, "Additional Comments Relating Watlings Island to San Salvador," #325; Charles A. Hoffman, "Archaeological Investigations at the Long Bay Site, San Salvador, Bahamas," #326; Robert H. Brill, I. L. Barnes, S. S. C. Tong, E. C. Joel, and M. J. Murtaugh, "Some European Artifacts Excavated on San Salvador Island," #323; Irving Rouse, "Origin and Development of the Indians Discovered by C," #285; John Winter, "San Salvador in 1492: Its Geography and Ecology," #486; Richard Rose, "Lucayan Lifeways at the Time of C.," #284; and Kathleen Deagan, "Arawak Responses to European Contact at the En Bas Saline Site, Haiti," #282.

[27] **Italy. Comitato Nazionale per le Celebrazioni del V Centenario della Scoperta dell'America.** *Nuova Raccolta Colombiana.* Rome: Istituto Poligrafico e Zecca dello Stato, 1988 –

A re-editing and amplification of the *Raccolta* of 1892-96 (#8), comprising texts, documents, essays, and monographs, with facing-page modern Italian translations of all items. The *Nuova Raccolta* specifically intends to make available to a general audience complete documentation of European activity related directly to and including the life and voyages of CC, reconciling scholarly accuracy with accessibility to a larger public than the *Raccolta* of 1892-96, which was addressed more specifically to specialists.

The titles and editors of the announced volumes (note that volumes 1, 6, and 11 were published in 1988 and are recorded in this Guide): 1. *Giornale di bordo,* ed. P.E. Taviani and Consuelo Varela, pub. 1988, #57; 2. *Relazione e lettere sul 2o, 3o, e 4o viaggio,* ed.

P.E. Taviani and Consuelo Varela. 3. *Scritti* (Miscellaneous writings by CC), ed. P.E. Taviani *et al.* 4. *I documenti genovesi e liguri,* ed. Aldo Agosto. 5. *Le scoperte de Cristoforo Colombo nelle relazioni sincrone di scrittori italiani,* ed. Gabriella Airaldi. 6. *La scoperta del Nuovo Mondo negli scritti di Pietro Martire d'Anghiera,* ed. Ernesto Lunardi et al., pub. 1988, #79. 7. *Le scoperte di Cristoforo Colombo nelle testimonianze di Chanca e di Bernáldez,* ed. Anna Unali. 8. *Le Historie di Don Fernando,* ed. P.E. Taviani and Francesca Cantù. 9. *Le scoperte di Cristoforo Colombo nei testi di Bartolomeo de Las Casas,* ed. Francesca Cantù. 10. *Le scoperte di Cristoforo Colombo nei testi de Fernández de Oviedo,* ed. Francesco Giunta. 11. *La Liguria e Genova al tempo di Colombo,* ed. Gaetano Ferro, pub. 1988, #28. 12. *Genova sul piano storico al tempo di Colombo,* ed. Geo Pistarino. 13. *Colombo e Chio e nel Mediterraneo,* ed. Geo Pistarino. 14. *La tradizione cartografica genovese e Colombo,* ed. Gaetano Ferro. 15. *La presenza degli italiani in Portogallo nel periodo Colombiano,* ed. Luisa D'Arienzo. 16. *Colombo in Spagna: setti anni decisivi della sua vita, 1485-1492,* by Juan Manzano Manzano, in Italian translation. Cf. #229. 17. *La presenza degli italiani in Spagna nel periodo colombiano,* ed. Luisa D'Arienzo. 18. *Le navi di Cristoforo Colombo,* ed. Cesare Ciano. 19. *Colombo nelle grandi opere letterarie,* ed. Giuseppe Bellini. 20. *Bartolomeo Colombo,* ed. Aldo Albonico. 21. *Paolo del Pozzo Toscanelli,* ed. Franco Cardini. 22. *Amerigo Vespucci,* ed. Ilaria Luzzana Caraci. 23. *Giovanni Caboto,* ed. M. Ballesteros Gaibrois. 24. *L'America come la vide Colombo,* ed. F. Moya Pons. 25. *Archeologia della scoperta colombiana,* ed. Veloz Maggiolo. 26. *Atlante con le rotte esplicative dell'Atlantico da Lanzarotto Malocello a Magellano,* ed. Osvaldo Baldacci. 27. *Iconografia Colombiana,* ed. Istituto della Enciclopedia Treccani.

[28] **Gaetano Ferro,** ed. *La Liguria e Genova al tempo di Colombo. Nuova Raccolta Colombiana,* Vol. 11. Rome: IPZS, 1988. 2 tomes.

Appendices, Maps, Bibliography, Index.

Tome 1, 283 pp. A portrait of the geographical and demographic conditions of Liguria and Genoa, 1450-1530. Synchronic essays on (1) the region in general, by Gaetano Ferro; (2) the natural surroundings, by Ferro; (3) the population and settlements, by Maria Pia Rota; (4) human activities, by Rota; (5) Genoa and vicinity, by Pietro Barozzi; (6) the western shoreline, by Rota and Ferro; (7) the eastern shoreline, by Daniela Galassi; (8) concluding inferences by Ferro. Documentary appendices by Rota and Stefanella Guarda. Bibliography by Guarda.

Tome 2. 18 maps illustrating the essays in tome 1.

II

Texts of Primary Documents

A. GENERAL

[29] **R. H. Major,** ed. *Select Letters of Christopher Columbus, with Other Original Documents, Relating to his Four Voyages to the New World.* London: Hakluyt Society, 1847. 240 pp.

Rpt. with new Introduction by John E. Fagg (New York: Corinth, 1961). An English translation of Navarrete, vol. 1 (#2), omitting the *Journal of the First Voyage,* several documents from the account of the fourth voyage, and CC's letters to his son Diego and Fr. Gorricio.

[30] **Benjamin Franklin Stevens,** ed. *Christopher Columbus: His Own Book of Privileges, 1502.* Intro. by Henry Harrisse. London: B. F. Stevens, 1893. 285 pp.

Index. Plates. A monumental edition, with facsimile reproduction of the handwritten Spanish text and, on each facing-page, transliteration in italic type and English translation in black letter, both by George F. Barwick. Full-page facsimile of illuminated title page and coat of arms. Historical intro. by Harrisse (pp. xiii-lxvi) in 9 chaps. describing the origin of four cartularies of the *Book,* and tracing the history of the three whose existence could then be identified, i.e., the Paris copy, the Genoese copy, and the "Boston" copy (then lost) purchased in Florence by Edward Everett in 1818. (Harrisse did not know of the Veragua copy, exhibited by the Duke of Veragua at the Chicago World's Fair in 1893.) The "Boston" copy was later purchased from *William* Everett by the LC, and has since been known as the "Washington" copy. See #89.

Stevens adds 3 pertinent letters from CC (pp. 267-75): two addressed to Niccolò Oderigo, the Genoese ambassador to Spain, and the third to the Bank of St. George, Genoa.

[31] **Fidel Fita,** ed. "Fray Bernaldo Buyl. Documentos inéditos," Bol R Acad Hist (Madrid), 22 (1893): 373-78.

Two previously unpublished royal documents. The first, a royal order by King Ferdinand, shows that Buyl in 1479 was secretary to

15

the king. Fita uses this to refute Roselly de Lorgues' denigration of Buyl.

[32] **L. T. Belgrano and M. Staglieno,** eds. *Il Codice dei privilegi di Cristoforo Colombo edito secondo i monoscritti di Genova, di Parigi, e di Providence. Raccolta,* II.2 (1894). 120 pp.

Textual preface, pp. ix-xix. Table of letters, privileges, and cedulas included, pp. 7-8. The documents, pp. 9-102. Pp. 105-114, text of the *Capitulations of Santa Fé.* A critical edition of the magnificently printed *Book of Privileges,* published by CC in 1494 *[sic],* based on the two extant copies in Paris and Genoa, with variant readings from a manuscript in the JCB compiled after the "1494" ed. had been printed.

[33] **Prospero Peragallo.** "Sussidi documentari per una monografia su Leone Pancaldo," *Raccolta* V.ii. (1894): 263-306.

Publishes 8 documents, including an archival record, a contract, a petition to Charles V, and several letters, to serve for a prospective monograph on Pancaldo, agent for CC's son Diego C.

[34] **L. T. Belgrano and M. Staglieno,** eds. *Documenti relativi a Cristoforo Colombo e alla sua famiglia. Raccolta,* II.2 (1896). 320 pp.

The Preface, pp. 1-80, introduces and discusses the documents, in this order: Those relating to (1) Giovanni C, CC's grandfather; (2) Domenico C, CC's father, to 1470; ; (3), Domenico C, 1470-72; (4) D.C., 1470-74, esp. in Savona; (5) CC in Spain; (6) CC back in Genoa; (7) CC the admiral of Spain; (8) Diego, CC's son; (9) Bartolomeo and Diego C, CC's brothers; (10) Fernando C, CC's 2nd son, and his mother Beatriz Enríquez; (11) María de Toledo, Diego's widow, and her descendants.

The documents follow, in order, with locations cited. All in original language, but translations, if any, are cited in the notes.

[35] **Julius E. Olson and Edward Gaylord Bourne,** eds. *The Northmen, Columbus, and Cabot, 985-1503.* New York: Scribners, 1903. 443 pp.

English translations. Under "The Northmen," Olson presents the Vinland narrations in the *Saga of Eric the Red* and in the *Flat Island Book* and extracts from Adam of Bremen and from the *Icelandic Annals,* plus relevant papal letters of Nicholas V (1448) and Alexander VI (1492). Under "Columbus," Bourne presents the *Capitulations,* the royal letter granting CC's title (30 Apr 1492), Markham's version of the *Journal of the First Voyage,* the letter to Santangel, Chanca's letter on the 2nd voyage, Las Casas' narrative of the third voyage, and CC's letter on the 4th voyage & other matters. Under "Cabot" Bourne presents the letters of Pasqualigo and Soncino (1497) and Pedro de Ayala's letter to the Cath. Sovereigns of 25 July 1498.

C. R. Benzley

[36] **Augustine M. Fernández de Ybarra,** trans. & ed. "The Letter of Dr. Diego Alvarez Chanca, Dated 1494, Relating to the Second Voyage of Columbus to America," Smithsonian Institution *Miscellaneous Collections,* 48, pt. 4 (1907): 428-57.

Very extensive explanatory notes and remarks on history and geography.

[37] **Vicente Vinau, M. Pérez Villamil, y J. Pérez de Guzman y Gallo.** "Documentos de Colón de la casa ducal de Veragua," *Bol R Acad Hist* (Madrid), 70 (1917): 126-30 & 468-71.

The Academia de la Historia's account and enumeration of the Duke of Veragua's hereditary collection of 97 priceless documents, largely by CC, the Catholic Monarchs, and the emperor Charles V. Both sections of the article appeal passionately to the Spanish government to pay the assessed price and acquire the documents rather than let them be sold abroad.

Note: the Spanish government's decision to purchase the documents from the Duke of Veragua was announced in the *New York Times,* 29 Oct 1925.

[38] **Angel de Altolaguirre.** "Algunos documentos inéditos relativos a don Cristóbal Colón y su familia," *Bol R Acad Hist* (Madrid), 92 (1928): 513-525.

The texts of 5 important documents: 1. A power of attorney given by CC to Diego and Fernando C's tutor Jerónimo de Aguero; 2. Another POA given by CC to Francisco Bardi in Salamanca, on 10 Dec 1505; 3. A POA given by Diego Colón on 4 Jun 1505 confirming CC's POA to Bardi on 10 Dec 1505; 4. Francisco Bardi's successful petition, based on Diego's POA of 4 Jun 1505, claiming 2000 pesos sent from the Indies for CC; 5. A POA authorized 8 Aug 1508 in Seville by Bartholomew Perestrello, CC's Portuguese brother-in law and uncle of Diego.

These documents establish that CC in the last months of 1505 was still moving with the court, not abed in Valladolid; that CC's payments of money from the Indies had not been cut off by the crown; and that Bartholomew Perestrello in 1508 was present in the same quarter of Seville as Diego.

[39] **Roberto Almagià.** *Gli Italiani i primi esploratori dell'America.* Rome: Libreria dello Stato, 1937. 515 pp.

A collection of reprints of important documents on CC, Vespucci, John and Sebastion Cabot, and Giovanni and Girala Verrazzano. Tables IV-VI following p. 458 contain a copy of the Assereto Document.

Rina Ferrarelli Provost

[40] **Ciriaco Pérez Bustamente,** ed. *Libro de los privilegios del Almirante don Cristóbal Colón (1498).* Madrid: Real Academia de la Historia, 1951. 203 pp.

A line-by-line transcription of the parent Veragua MS (1498) of CC's *Book of Privileges.* Preliminary study, pp. ix-lxv. The transcription is accompanied by variant readings from the derivative, printed Paris and Genoa versions. Variants in the Florentine (Washington, LC) and Providence (JCB) copies are not indicated, perhaps because the Florentine is not considered a careful copy and the Providence is a fragmentary MS separately compiled.

Includes photo-facsimile of the Veragua MS. Index of documents, names, and places. For filiation, see H. Putnam, #88, F. G. Davenport, #89, and C. L. Nichols, #92.

[41] **Jaime Colomer Montset.** "Las Capitulaciones de Santa Fé registradas en el archivo de la Corona de Aragón en Barcelona," *Studi Colombiani* (Genoa: SAGA, 1952), 2: 391-97.

A transcription of the document, with textual notes and an introduction emphasizing the curious fact that, though registered in Aragón, the Capitulations do not mention Aragón but only Castile and León. Other points: (1) Ferdinand and Isabel do not enter the agreement for Aragón and Castile, but personally. This perhaps acknowledges Portugal's resistance to Aragón as an Atlantic power. (2) The wording strongly suggests (to the author at least) that CC had already been to the New World and knew exactly where he was going. (3) Unless the sovereigns had been convinced that CC knew what he would find and where it was, the enormous concessions would have made no sense.

[42] **Ernst Gerhard Jacob.** *Christoph Columbus: Bordbuch, Briefe, Berichte, Dokumente.* Bremen: Schönemann, 1956. 450 pp.

An extensive, annotated selection of documents recording the life and death of CC, all translated into German. Excellent introduction, pp. 13-23: a compendium of information placing CC's life in its context, including a useful dictionary of Columbiana, pp. 405-450. No textual information on translations.

[43] **Aldo Agosto.** "Due nuovi documenti Colombiani dell'archivio di stato di Genova," *Atti II Conv Internaz Stud Col* (Genoa: CIC, 1977), pp. 89-101.

Describes, analyzes, and publishes two legal documents of 9 Aug 1469 and 20 Nov 1469, respectively, which confirm prior data recording (1) the presentation of a difference of opinion (before notary arbiters) about a house left by Domenico C. to Guglielmo Fontanarossa (probably a cousin of Domenico's wife Susanna), and (2) the resolution of the quarrel, the parties having agreed beforehand to abide by the arbiters' decision. These were Bernardo Camogli and Giovanni Agostino Guano. They both found in favor of D.C., agreeing with him that G. Fontanarossa owed him 10 Genoese lire in back rent and, in addition, they made the contract more specific.

[44] **Consuelo Varela,** ed. *Cristóbal Colón: Los cuatro viajes. Testamento.* Madrid: Alianza, 1986. 304 pp.

A new edition drawing together the documents narrating CC's 4 voyages, along with CC's will of 19 May 1506, recorded by Pedro de Enoxedo, and the codicil recorded by Pedro de Azcoitia.

Up-to-date introduction, pp. 9-36, places the material in historical context, and keys it to two other editions by Varela, *CC: textos y documentos completos,* #54, and *Cartas de particulares a C. y relaciones coetaneas* (with Juan Gil), #73.

[45] **Carlos Sanz,** ed. *Confirmación de los privilegios que los Reyes Católicos hicieron en favor de don Cristóbal Colón.* No place: Carlos Sanz, no date. 4pp.

The only known printed copy of this document is preserved in the Ducal Library of Alba, in Madrid.

B. THE WRITINGS OF CHRISTOPHER COLUMBUS

[46] **Cesare de Lollis,** ed. *I Scritti di Cristoforo Colombo. Raccolta,* I. Rome: Ministro della Pubblica Istruzione, 1892-94.

To date, the only critical edition of all the known writings with full apparatus.

Contents. Part I, Vol. 1 (1892): (1) *Journal* of the First Voyage; (2) *Letter* to Santangel and Sánchez; (3) CC's *Memorandum* of the Second Voyage; (4) *Journal* of the 2nd voyage, reconstructed from passages in Ferdinand Columbus's *Historie* and Las Casas' *Historia de las Indias;* (5) CC's *Instructions* to Antonio de Torres (6) CC's *Instructions* to Pedro Margarite; (7) Fragment of a letter to the Catholic Monarchs; (8) CC's *Memorandum* of the 3rd Voyage (9) Fragment of a memorandum of the 3rd Voyage; (10) Contract of CC and Fonseca with Anton Marino; (11) CC's *Letter* to the Bishop of Badajoz; (12) Fragments of a letter from CC to his brother Bartholomew; (13) a receipt from CC; (14) CC's will of 22 Feb 1498, including the institution of the Mayorazgo. In the front matter appears a critical introduction to each document.

Pt. I, Vol. 2 (1894), in 2 parts: pt. 1, pp. 1-160; pt. 2, 161-570. (15) CC's 3rd voyage, from Las Casas' *Historia;* (16) CC's account of the 3rd voyage; (17-22) surviving excerpts from 6 letters to the monarchs; (23) *Letter* to Francisco Roldán; (24) safe conduct for Roldán; (25-26) fragments of letters to the monarchs; (27) privilege in favor of Pedro de Salcedo; (28-29) fragments of letters to the monarchs; (30) letter to some persons of the court; (31) *Letter to Prince Juan's Nurse* (doña María de Torres) (32) the *Book of Prophecies* (33) letter to the monarchs; (34) letter to Pope Alexander VI; (35) letter to Niccolò Oderigo; (36) memorandum to his son Diego; (37) letter to the Bank of St. George; (38-40) letters to Fr. Gorricio; (41) CC's account of the 4th voyage; (42) fragment of a letter to Ovando; (43) letter to Ovando; (44-46) orders for payments to Diego Rodriguez, Rodrigo Viscáyno, and Diego de Salcedo, resp.; (47-52) letters to Diego; (53) letter to Niccolò Oderigo; (54) letter to Diego; (55) letter to Fr. Gorricio; (56-58) letters to Diego; (59-60) fragments of letters to King Ferdinand; (61) from a memorandum for K. Ferdinand; (62) from a letter to Diego de Deza; (63) from a memorandum for K. Ferdinand; (64) CC's *Testament* of 19 May 1506.

Pt. I, Vol. 2, pt. 2. Postils to CC's Books: (1) to Pius II's *Historia Rerum,* etc.; (2) to treatises of d'Ailly; (3) to the *Book of Marco Polo;* (4) to Pliny's *Naturalis Historia;* (5) to Plutarch's *Parallel Lives;* (6) to Ptolemy. Index to Part I, Vols. 1-2, pp. 529-70.

Pt. I, Vol. 3 (1892). Cesare de Lollis, ed., *Autografi di Cristoforo Colombo con Prefazione e Trascrizione Diplomatica.* Photographs, with facing-page transcriptions, of the writings published in Vol. I, parts 1 & 2.

[47] **Antonio Ballesteros Beretta.** "Una Carta Inédita de Cristóbal Colón," *Rev Indias* (Madrid), 9 (1949): 489-506.

A transcription of a previously unpublished letter of CC written from Seville on 28 Dec 1504 to an unknown addressee in the admiral's formal handwriting, which he reserved for important personages. The contents reflect CC's agitation, at the time of his return from the 4th voyage, over the question whether he could fully establish the privileges granted in the *Capitulations* and pass them on to his son Diego. Ballesteros reviews in scholarly fashion pertinent aspects of the manuscript: the text, the context within which it was written, and the question of its authenticity, of which he is convinced.

[48] **Antonio Muro Orejón.** "Un autógrafo de Cristóbal Colón," *Studi Colombiani* (Genoa: SAGA, 1952), 2: 127-35.

Describes and publishes an "albalá" or elaborate receipt, signed by CC, for 70,000 maravedis from the royal funds designated for the preparation of the fleet for the second voyage. The document is preserved in the Archive of Notarial Protocols in Seville: Oficio IX Escribanía de Luis García de Celada. Legajo primo de 1498. Folio 158 verso.

[49] **Cecil Jane,** tr. The *Journal of Christopher Columbus,* rev. and ann. by Louis-André Vigneras. New York: Potter, 1960. 227 pp.

Illus. Notes. Bibliography. Critical foreword by Vigneras, xv-xxiii. Appendix by R. A. Skelton. This edition seems to be a response to S. E. Morison's textual critique of the first edition (1930) in *Hisp Amer Hist Rev,* 19 (1939), 235-61, #95; but the notes by Vigneras are explanatory, not textual. All of V's textual corrections are silent, not mentioned in the notes.

Includes a translation of CC's letter on the 1st voyage. The Skelton appendix discusses the cartography of CC's 1st voyage, and describes how world maps were compiled. Adds comments on Henricus Martellus, on Almagià's Columbian prototype, and on the Juan de la Cosa and Cantino voyages.

[50] **Alexandre Cioranescu,** ed. and trans. *Oeuvres de Cristophe Colomb.* Paris: Gallimard, 1961. 527 pp.

French translation of the writings available at the time.

[51] **Joaquín Arce Fernández and Gil Esteve,** eds. *Diario de a bordo de Cristóbal Colón.* Alpignano, Italy: Tallone, 1971. 188 pp.

Introductory essay, "Significado lingüístico-cultural del diario de Colón," pp. 11-28, by Arce. Sections: (1) CC, figura problematica; (2) El diario del descubrimiento; (3) Maravilla e hipérbole; (4) Aspectos estilísticos; (5) Léxico, formas, construcciones; (6) De la lengua a la literatura. A thorough analysis of the style of the Diario: diction, hyperbolic; figurative language, simile predominating; grammar and syntax, idiosyncratic and sometimes unidiomatic, but more likely to show Italian or marinesco influence than

Portuguese; and despite Las Casas' interference, highly spontaneous and direct.

Text based on a new transcription by M. Gil Esteve of the Las Casas original MS in the Biblioteca Nacional of Madrid.

[52] **Antonio Rumeu de Armas.** *Un escrito desconocido de Cristóbal Colón: el memorial de la mejorada.* Madrid: Cultura Hispánica, 1972. 86 pp.

Publishes, analyzes, and comments on CC's July 1497 memorandum to the crown interpreting the Alexandrine bulls and the treaty of Tordesillas dividing the areas to be exploited by Spain and Portugal. Points out what CC considers fundamental violations of the treaty by Portugal. Intended for royal use in negotiating with the Portuguese.

[53] **Aldo Agosto.** "La scoperta nel 1610 della lettera di Colombo ai protettori delle compere di San Giorgio," *Atti III Conv Internaz Stud Col* (Genoa: CIC, 1979), pp. 521-36.

Publishes the exultant letter written by Gov. Agostino Sivori to the bank officers on 5 May 1610 reporting his discovery, and another of 30 Apr 1612 to the same, concerning a bequest of CC claimed by some of his descendants.

[54] **Consuelo Varela,** ed. *Cristóbal Colón: Textos y Documentos Completos; Relaciones de Viajes, Cartas y Memoriales.* Madrid: Alianza, 1982. 353pp. 2nd rev. ed., 1984, 389pp. 3rd rev. ed., 1986.

Prologue and notes by Varela. Glossary, Indexes, Maps. The 93 documents in the 2nd ed. surpass by 29 the most complete previous collection. The excellent prologue contains most of the elements of the textual preface to a fully critical edition, with attention to (a) the provenance of those texts that exist in more than one substantive edition or MS; (b) the fidelity of available copies where CC's holographs are not extant; (c) techniques for establishing the most probable reading of a doubtful locution; (d) the significance for the textual scholar of wide and deep knowledge of the milieu. No textual notes, however, and no specific account of the choice of copy-texts for those documents (like the "Letter to Santangel") that exist in more than one substantive form.

Documents in 2nd ed.: 1. Apostilas: 17 postils on these topics: Ireland, La Mina, CC in Lisbon after 1485, the "papers" of CC, receipts; 2. *Diario,* 1492-93; 3. Fragmento de un escrito en el Cuaderno de a Bordo, n.d.; 4. Carta a Rodrigo de Escobedo, 4Jan93; 5. Carta a Luis de Santangel, 4Mar93; 6. Ordenanza de Colón, 20Feb93; 7. Memorial de A. Torres, 30Jan94; 8. Instrucción a Mosén

Pedro Margarite, 4Apr94; 9. Fragmento de una carta a los reyes, Jan95; 10. Nombramiento de Teniente de Gobernador a don Bartolomé Colón, 17Feb96; 11. Memorial de la Mejorada, Jul97; 12. Poder a Jerónimo de Aguero, 31Oct97; 13. Memorial a los Reyes sobre la población de las Indias, n.d.; 14. Fragmento de un Memorial a los Reyes, n.d.; 15. Contrato de Cristóbal Colón y Fonseca con Antón Marino, 1Jan98; 16. Carta a don Juan de Fonseca, Obispo de Badajoz, before 23Jan98; 17.Conocimiento de deuda, 17Feb98; 18. Fragmentos de una carta a Bartolomé Colón, Feb98; 19.Institución de Mayorazgo, 22Feb98; 20. Albalá a Ximeno de Briviesca, 26Apr98; 21. Carta a don Diego Colón, 29Apr98; 22-23. Cartas a fray Gaspar de Gorricio, 12&28May98; 24. *Relacción del Tercer Viaje*, 1498; 25-31. Fragmentos de cartas a los Reyes, 1498-1500; 32. Carta a Francisco Roldán, 20Oct98; 33. Salvoconducto a Francisco Roldán, 26Oct98; 34. Relación de la gente que fue con Cristóbal Colón en el Primer Viaje, 16Nov98; 35. Carta a Miguel Ballaster, 21May99; 36-37. Fragmentos de cartas a los Reyes, n.d.; 38. Provisión a Pedro Salcedo, 3Aug99; 39. Fragmentos de una carta a los Reyes, 10Sept99; 40. Traslado de una carta de los Reyes a Bobadilla con la respuesta del Almirante 30May-15Sept1500; 41. Carta a doña Juana de Torres, ama del Principe don Juan, n.d.; 42. Hoja suelta en papel de mano del Almirante, ca.1500; 43. Memorial a los Reyes, n.d.; 44. Memorial preparatorio del Cuarto Viaje, n.d.; 45. Carta a los Reyes, 1501; 46-51. Cartas a fray Gaspar de Gorricio, 26Feb,14&24May,9Jun,13Sept1501,&n.d.; 52. *Libro de las profecías*, n.d; 53. Información de los Privilegios y Mercedes, n.d.; 54. Respuesta del Almirante a los capítulos de sus Privilegios, n.d.; 55. Memorial de agravios, n.d.; 56. Memorial del Almirante sobre agravios que recibió, n.d; 57. Carta a la Reina, Aug-Sept1501; 58. Conocimiento de cien castellanos de oro a Alonso de Morales, 22Oct1501; 59. Carta a los Reyes, 6Feb02; 60. Memorial a Diego Colón antes de emprender el Cuarto Viaje, n.d.; 61. Carta al Papa Alejandro VI, Feb02; 62. Carta a Nicolás Oderigo, 21Mar02; 63. Carta a la Banca de San Jorge, 2Apr02; 64-65. Cartas a fray Gaspar de Gorricio, 4Apr02, 20-25May02; 66. *Relación del Cuarto Viaje*, 7Jul03; 67. Carta a fray Gaspar de Gorricio, 7Jul03; 68-69. Cartas a Nicolás de Ovando, Mar04, 3Aug04; 70. Libramiento a favor de Diego Rodríguez, 7Sept04; 71. Libramiento a favor de Rodrigo Vizcáino y de Francisco Niño, 8Sept04; 72. Libramiento a favor de Diego de Salcedo, 9Sept04; 73-79. Cartas a Diego Colón, 21Nov,28Nov,1Dec,3Dec,13Dec,21Dec04; 80. Carta a Nicolás Oderigo, 27Dec04; 81. Carta a Juan Luis de Mayo 27Dec04; 82. Carta a Diego Colón 29Dec04; 83. Carta a fray Gaspar de Gorricio, 4Jan05; 84-86. Cartas a Diego Colón, 18Jan-5Feb05; 87. Carta al Rey don Fernando, Jun05; 88. Poder a favor de Francisco de Bardi, 10Dec05; 89-90. Fragmentos de una carta y un memorial al Rey don

Fernando, n.d.; 91. Fragmento de una carta a fray Diego de Deza, n.d.; 92. Carta a los Reyes don Felipe y doña Juana, Spring 06; 93. Testamento, 19May06.

[55] **Oliver Dunn,** ed. "The Diario, or Journal, of Columbus's First Voyage: a New Transcription of the Las Casas Manuscript for the Period October 10 through December 6, 1492," *Terr Incog,* 15 (1983): 173-231.

Rpt. in *In the Wake of Columbus,* #19, pp. 173-231. The transcription is preceded by 5 paragraphs explaining the principles employed. Attempts "to present a text as close to that of the manuscript as can be produced on an ordinary typewriter equipped with an almost standard keyboard". Excerpt from edition of the complete transcript, due 1989 from Univ. of Oklahoma Press.

[56] **Gaetano Ferro** trans. & ed. *Diario de bordo: Libro della prima navigazione e scoperta delle Indie,* by Cristoforo Colombo. Milan: Mursia, 1985. 237 pp.

A translation of the M. Gil Esteve transcription of the Las Casas MS in the Biblioteca Nacional in Madrid. Introduction, translation and explanatory notes by Ferro.

[57] **Paolo Emilio Taviani and Consuelo Varela,** eds. *Il Giornale di Bordo, Libro della Prima Navigazione e Scoperta delle Indie. Nuova Raccolta Colombiana,* I. Rome: IPZS, 1988. 2 tomes.

Tome 1, 321 pp, contains (a) Varela's transcription of the Spanish text of CC's *Journal of the First Voyage* as abstracted in Las Casas' MS in the Biblioteca Nacional in Madrid; (b) the de Lollis text (1892) of CC's "Letter to Luis de Santangel" of 4 Mar 1493. Both texts are accompanied by facing-page Italian translations by Taviani with the collaboration of Marina Conti and Francesca Cantù.

Tome 2, 515 pp., contains (a) Varela's apparatus for her transcription of the *Journal,* comprising ten essays; (b) Varela's apparatus for the "Letter to Santangel," comprising two essays; (c) Taviani's historico-geographic reconstruction of the first voyage, comprising 78 essays; (d) a bibliography of the editions of the *Journal,* 1825-1987, by Simonetta Conti; (e) three indexes, of persons, places, and bibliographical references. Facing p. 208 is a chart of the first voyage.

C. EARLY BIOGRAPHIES AND OTHER PRIMARY DOCUMENTS NOT BY CC HIMSELF

[58] **Andrés Bernáldez.** *Historia de los Reyes Católicos D. Fernando y Da. Isabel.* Granada: Zamora, 1856. 2 vols.

Chapters 119-34, vol. 1, pp. 278-334, constitute a basic primary source for the 2nd voyage.

Rina Ferrarelli Provost

[58a] **Gonzalo Fernández de Oviedo y Valdés.** *Historia general y natural de las Indias.* Madrid: Real Academia de la Historia, 1851-55. 4 vols.

Rpt., with preliminary study by Juan Pérez de Tudela y Bueso, *Biblioteca de autores espannoles* (Madrid: Atlas, 1959), Vols. 117-121. Chaps. 1-9 of Book 1, pp. 10-80 in Vol. 1 of the 1851 edition, constitute a basic primary source for CC's life and voyages, ranking with Peter Martyr's *Decades* (see #79; English trans., #63), FC's *Historie* (see #66; English transl, #71); and Las Casas' *Historia* (see #59 and #67).

Book 1, containing the life of CC, was published in 1535; the other three volumes were not published until the 1851-55 edition.

[59] **Bartolomé de Las Casas.** *Historia de las Indias* (Madrid: Ginesta, 1875-76). 5 vols.

1st publication of the history of the Spanish occupation of the Western hemisphere through 1520, written over the period ca. 1527-1561. Includes the most extensive and detailed contemporary biography of CC, by the man who had the most complete opportunity to know the circumstances and the ambience of CC's life. CC treated in 1: 41-501 (Bk.I,chaps. 2-82); 2 throughout (chaps. 88-183); and 3: 1-514, (Bk.II, chaps. 1-38). Appendix, 5: 237-555, describes Hispaniola and "the Indies" generally. Probably the best source for CC's life; critical ed. #67. English trans. of selected passages on CC, A.M. Collard, ed., *History of the Indies* (New York: Harper, 1971), 302 pp. Harper Torchbook #s TB1540 and TLE131.

[60] **L. T. Belgrano,** ed. "Lettera del Re Emanuele di Portogallo a Fernando e Isabel," *Boll Soc Geog Ital* (Rome), Ser. 3, 3 (1890): 271-87.

Title continued: "di Castiglia, sopra la navigazione di Pedro Alvarez Cabral, nel suo ritorno dal Brasile, alla costa d'Africa (1500-1501)." Account of the discovery of Brazil. Publishes in parallel columns (1) the previously unknown original Portuguese text from a copy made by a Venetian representative at the court of Portugal, Giovanni Pasqualago, and (2) the synchronous Spanish text published by Navarrete, #1, tome III, no. 13, pp. 94-101.

[61] **Cesáreo Fernández Duro,** ed. *De los pleitos de Colón.* Madrid: Sucesores de Rivadeneyra, 1892-94. 2 vols. *Colección de documentos inéditos . . . de Indias* (#4), 2nd series, vols. 7-8.

Vol. 1: editor's intro., i-xxiii. Docs. 1-56, 1506-1514. 3 indexes: docs (chron.), persons, and places.

Vol. 2: editor's intro., i-xv. Docs 57-225, 1497-1514. 3 indexes: docs (chron.) persons, and places.

1st vol. records, chronologically, documents and testimony introduced in court in the first of the Colóns' long series of suits agains the Spanish crown claiming the privileges granted to CC in 1492. Intro. succinctly recounts progress and disposition of the suits—first against the crown and then between factions of the family—over 3 centuries, 1508-1796.

2nd vol. cites various newly published documents (57-78) supplementing the first vol.; then records docs and testimony in two subsequent suits appealing the successive judgments of 1511 and 1520, appeals that led to a 3rd judgment in 1525. Both vols. record only by reference the documents that are readily available elsewhere.

For summary note on "Los Pleitos," see Otto Schoenrich, *The Legacy of Colón,* Introduction. (#619).

[62] **C. Errera.** "Un nuovo documento colombiano," *Riv Geog* (Florence), 7 (1910): 600-601.

Brief account of a letter dated 4 Aug 1496 from a Florentine, Jacopo Acciaioli, at Ferrara, referring to a fragment of a letter from the Genoese Francesco Cattaneo, dated 11 Jun 1496, from Cádiz. Cattaneo describes CC's arrival at Cádiz with two caravels (end of 2nd voyage). CC showed a map detailing his 10,000 mile voyage, and reported discovering an island 3 times as large as England. Cattaneo took a number of American natives to shore in his boat; these included three ferocious female cannibals.

Errera reports that the dispatch from Acciaioli of 4 Aug 1496 was published in *Archivio Storico Italiano* in Aug. 1910.

[63] **Francis Augustus MacNutt,** ed. & trans. *De Orbo Novo, the Eight Decades of Peter Martyr d'Anghera.* New York: G. P. Putnam, 1912. 2 vols.

English translation of Peter Martyr's Decades: *From the New World,* with Introduction, Notes, and Bibliography. Vol. 1, Decades 1-3; Vol. 2, Decades 4-8.

[64] **Alicia Bache Gould,** ed. "Documentos inéditos sobre hidalguía y genealogía de la familia Pinzón," *Bol R Acad Hist* (Madrid), 91 (1927): 319-75.

7 documents, 1501-1777, beginning with the order from King Ferdinand knighting Vicente Yañez Pinzón.

[65] **Edmund Buron,** ed. *Imago Mundi de Pierre d'Ailly.* Paris: Maisonnueve, 1930.

A facing-page French translation of the 4 cosmographical treatises that D'Ailly's *Imago Mundi* comprises, along with the marginal notes from CC's personal copy in the Colombina Library in Seville. Introductory essay on (a) the influence of D'Ailly on CC and (b) the life and works of D'Ailly. See #761.

[66] **Rinaldo Caddeo,** ed. *Le historie della vita e dei fatti di Cristoforo Colombo,* by Ferdinand Columbus. Milano: Alpes, 1930. 2 vols.

Introductory essay, I,iii-lxviii. After an account of F's life, Caddeo defends the authenticity of F's authorship and the general truthfulness and authority of the *Historie,* while admitting F's tendency to cover up CC's humble origin and lack of formal education. This defense constitutes an influential turning point, a reaction against the tendency to distrust the *Historie* begun by Harrisse in his 1871 study *D. Fernando Colón* (#757), a tendency that dominates CC scholarship well into the 1930's. The introduction concludes with an explanation of the editorial policy respecting the text (not a rigorous critical edition). Bibliography of previous editions and of CC studies, I,lxxi-lxxvii.

The Text, I, 3-316 and II, 1-308. Appendices: 10 useful essays (1) on CC's Genoese origins & on competing non-Genoese claims; (2) CC's early Mediterranean voyages; (3) the French mariners Coullon and CC's arrival in Portugal; (4) voyages to Iceland (1477) and Genoa (1479); (5) CC, Toscanelli, and the birth of the Enterprise of the Indies; (6) Italian financiers in Spain and the financing of the voyages; (7) M.A. Pinzón; (8) Bartholomew C. and Bartholomew Dias; (9) Name and signature of CC; (10) a moral and physical portrait of CC.

[67] **Bartolomé de las Casas.** *Historia de las Indias,* ed. Agustín Millares Carlo; Introduction by Lewis Hanke. Mexico City and Buenos Aires: Fondo de Cultura Económica, 1951. 3 vols.

2nd ed. revised, 1972. The text, though termed "fide digne" by Lewis Hanke in his elaborate introductory essay, lacks textual notes and thus should probably be styled a "transcription" rather than a critical edition. It is, however, the nearest known approach to a critical edition of the *Historia*. Hanke's edition discusses Las Casas himself, his works, his sources, his status as a historian; the composition of the *Historia*; the fortunes of the MS; and the consequent problems facing the critical editor, Agustín Millares.

[68] **Montserrat Freixa Ubach,** ed. & trans. "El pasaport y la carta de presentación que los Reyes Católicos entregaron a Cristóbal Colón en 1492," *Studi Colombiani* (Genoa: SAGA, 1952), 2: 363-67.

Cites as an index of the lack of Aragonese participation in the Discovery the fact that only 3 fundamental CC docs were registered at the Aragonese court, viz., the *Capitulations*, CC's passport, and the letter of presentation. Presents Spanish translations of the passport and the letter, with commentary.

[69] **R. Konetzke,** ed. *Colección de documentos para la historia de la formación social de Hispanoamérica*. Madrid: Consejo Superior de Investigaciones Científicas, 1953. Vol. 1. 671 pp.

Items 1-29, pp. 1-63, 29 letters from the crown dated sequentially from 29 May 1493 to 5 Feb 1515, directed mostly to CC and Diego C., bearing orders establishing royal policy toward the native Americans.

[70] **Juan Pérez de Tudela Bueso,** ed. "Una Rectification y Tres Documentos," *Rev Indias* (Madrid), 13 (1953): 609-23.

(1) Corrects, by means of a document in the Simancas register, Navarrete's significant misreading of *dos* as *doce* in his transcription of the sovereigns' letter of 30 Apr 1492 giving CC command of the entire fleet of the 1st voyage; (2) prints a hitherto unpublished memorial in the Archive of Simancas of items and expenses for the 4th voyage; (3) (a) prints a memorial authorizing the purchase of the 4 ships for the 4th voyage and the various other things needed (both this and the first document show the remarkable degree to which the sovereigns accommodated CC's extreme demands); (b) the second part of the same memorial demonstrates—the author feels it is the first documentation not coming from CC himself—the existence of a group at court opposing CC's plans and proposals.

[71] **Ferdinand Columbus.** *The Life of Admiral Christopher Columbus by his Son Ferdinand,* translated and annotated by Benjamin Keen. New Brunswick NJ: Rutgers Univ. Press, 1959. 316 pp.

English translation based on Rinaldo Caddeo's 1930 edition (#66) of Ferdinand C's *Historie* (1571) and Ramón Iglesia's 1947 Spanish translation. Preface, pp. v-xviii, discusses the translator's principles, reviews FC's life & his virtues and limitations as a biographer, and discusses previous editions and translations.

[72] **Antonio Muro Orejón,** ed. *Pleitos Colombinos.* Seville: Escuela de Estudios Hispanoamericanos, 1967 – .

A modern critical edition. For summary chapter on "Los Pleitos," the litigation against the Spanish crown by CC's heirs to claim the rights granted CC in the Capitulations of 1492, see entry on Otto Schoenrich, *The Legacy of Columbus,* Introduction (#619).

Vol. 1. *Proceso hasta la sentencia de Sevilla (1511),* ed. A. Muro Orejón, Florentino Pérez Embid, and Francisco Morales Padrón. 1967. 234 pp.

General introduction to the series, pp. xv-xxvi. Intro. to v.1, xxix-xxxvi.

Vol. 2. *Pleito sobre el Darien (1512-1519),* ed. A. Muro Orejón. 1983. 158 pp.

Intro., pp. xv-xxxii.

Vol. 3. *Probanzas del almirante de las Indias (1512-1515),* ed. A. Muro Orejón. 1984. 458 pp.

Intro., pp. xi-l.

Vol. 8. *Rollo del processo sobre la apelación de la sentencia de Dueñas (1534-1536),* ed. A. Muro Orejón, F. Pérez-Embid, and F. Morales Padrón. 1964.

Intro., pp. xvii-xxxi.

[73] **Juan Gil and Consuelo Varela,** eds. *Cartas de Particulares a Colón y Relaciones Coetáneas.* Madrid: Alianza, 1984. 389 pp.

Indexes. Transcription and edition of 36 important documents providing accounts of CC's life and voyages and context for his letters and the events of his life. Designed to accompany #54.

[74] **Luis Arranz Márquez.** "Cuatro Documentos Colombinos," *Rev Indias* (Madrid), 45 (1985), 349-71.

Publishes and transcribes four documents concerning Diego Columbus, hitherto known only indirectly. The first locates Diego in Seville in 1508 and sets forth his refusal to accept an economic solution to his lawsuit against the crown. In the second (1511), Fernando Columbus proposes in Diego's name a circumnavigation of the world. The third and fourth (1512) are letters from Diego to Cardinal Cisneros about Hispaniola, reporting, i.a., a great mortality from colds. The first document is in the archive of the Duke of Alba, the others in the New York Public Library.

[75] **Luigi Giovannini,** trans. & ed. Marco Polo, *Il Milione: con le postille di Cristoforo Colombo.* Rome: Edizioni Paoline, 1985. 309 pp.

An Italian translation of the copy of the 1485 Latin edition of the *Book of Marco Polo,* with CC's notes, also translated into Italian, at the foot of each page. Historical and textual introduction, pp. 5-45, provides useful background.

[76] **Antonio Rumeu de Armas,** ed. "Un documento inédito sobre Cristoforo Colombo," *Saggi e Documenti* (Genoa), 6 (1985): 435-49.

Argues that the "limosna" to an unnamed Portuguese for 30 Castilian doblas, given in the accompanying, hitherto unpublished document at Linares on 18 Oct 1487 and signed by Pedro de Toledo, was given to CC to cover his needs at the time and enable him to return to Córdoba. CC was passing as a Portuguese not because he was one, but because this identity enhanced his image as a navigator, explorer, and cartographer.

[77] **Juan Gil,** ed. & trans. *El libro de Marco Polo: ejemplar anotado por Cristóbal Colón y que se conserva en la biblioteca capitular y colombina de Sevilla.* Madrid: Testimonio Compañía Editorial, 1986. 469 pp.

Elaborately annotated scholarly Spanish translation of the 1485 Latin edition of Marco Polo possessed by CC, with concomitant Spanish translation of CC's marginal notes. Preceded by important introductory essay that draws together with great cogency the various studies which tend to show that CC had not read or annotated Marco Polo by 1492 (he appears to have received his copy from John Day in 1497) and that CC's diligent annotation of his books was in reaction to the pressures created by his discovery and his various claims about it.

[78] **Juan Gil,** ed. & trans. *El libro de Marco Polo anotado por Cristóbal Colón. El libro de Marco Polo: version de Rodrigo de Santiella.* Madrid: Alianza, 1987. 287 pp.

Combines two editions of *The Book of Marco Polo* that are fundamental to the history of the Spanish discoveries: (1) the first Latin edition of 1485, the one owned and annotated by CC: both the Latin text and CC's notes are translated into Spanish; (2) the first Castilian translation of Marco Polo (Seville, 1519), made by Rodrigo Fernández de Santaella from the 1503 Venetian edition of the Latin text.

Gil's introductory essay (pp. i-lxix) places all this within the context of the Seville that CC and his friends and collaborators inhabited at the turn of the 16th c. This sheds important light both on the culture of the time and on CC himself. Not the least important element is Gil's conclusion (p. viii), already set forth in the introduction to his 1986 edition of Marco Polo, #77, that CC did not own or annotate Marco Polo's book until 1497.

[79] **Ernesto Lunardi, Elisa Magioncalda, and Rosanna Mazzacane,** eds. *La scoperta del Nuovo Mondo negli scritti di Pietro Martire d'Anghiera. Nuova Raccolta Colombiana*, VI. Rome: IPZS, 1988. 503 pp.

Textual appendix by Magioncalda and Mazzacane. Bibliography and Indexes of Names and Places by Simonetta Conti. A critical edition with textual introduction and notes, abstracts, and facing-page Italian translations by Magioncalda and Mazzacane. Five learned essays or "schede" by Ernesto Lunardi, viz., "Pietro Martire d'Anghiera: un uomo nelle tempeste della storia," pp. 369-392; "Lineamenti del pensiero di Pietro Martire," pp. 393-404; "Pietro Martire e Cristoforo Colombo," pp. 405-414; "L'Opus Epistolarum e il nuovo mondo," pp. 415-433; *"Le Decades de Orbe Novo,"* pp. 415-446.

III

Studies of Primary Documents

A. GENERAL: STUDIES OF VARIOUS DOCUMENTS
RELATED TO COLUMBUS

[80] **Alejandro María de Arriola.** "Los libros de Colón," *Boletín de la Sociedad Geográfica de Madrid*, 27 (1889): 272-86.

A report to the society describing three books in the Colombina library containing writing in CC's own hand: (1) Works by D'Ailly, including *Imago Mundi*; (2) Pius II's *Historia rerum*; (3) CC's MS *Book of Prophecies*. Arriola notes two other books which in his opinion belonged to CC, but were not annotated by him: *Marci Pauli de Veneciis consuetudinis et conditionibus orientalium regionum*, and another volume containing both Peter Martyr's *Decades of the New World* and a poem, *Legatio babilonica oceana deacs*.

None of the other books now conceded to contain notes by CC, e.g., the *Book of Marco Polo*, Pliny's *Naturalis Historia*, Plutarch's *Parallel Lives*, Ptolemy, are listed by Arriola.

[81] **John Boyd Thacher.** *Christopher Columbus: His Life, His Work, His Remains, as Revealed by Original Printed and Manuscript Records together with an Essay on Peter Martyr of Anghera and Bartolomé de las Casas, the First Historians of America.* New York and London: Knickerbocker, 1903. 3 vols.

Illus., Plates, Charts, Appendix. A study of the copious material on CC made available by the end of the 19th century. In 10 parts: (1) essays on life & works of Peter Martyr and Las Casas; (2) Intro. to treatment of CC; (3) CC the man; (4) the purpose; (5) the event; (6) the announcement; (7) exploration (voyages 2-4); (8) personality (studies of portraits and statues, handwriting, and Ferdinand Columbus and his library); (9) the fortunes of CC's mortal remains; (10) family tree of CC's descendants. Separate entries in this Guide for the essays on portraits (#683), handwriting (#452), and mortal remains (#687).

A monumental project of documentation, translation, and analysis, occasionally lacking in scholarly method but extremely useful.

[82] **M. Serrano Sanz.** "El archivo colombino de la Cartuja de las Cuevas," *Bol R Acad Hist* (Madrid), 97 (1930), 145-256, 534-637.

Describes the monastery of Las Cuevas in Seville and enumerates the holdings of the Columbian archive.

[83] **William Jerome Wilson.** "The Textual Relations of the Thacher Manuscript on Columbus and Early Portuguese Navigations," *Papers Bib Soc Amer* (Portland ME), 34 (1940): 199-220.

Comparative textual study which concludes that the Thacher MS was copied from letters and MSS sent to Venice by Angelo Trevisan prior to 1503.

[84] **Giovanni Praticò.** "La scoperta dell'America nei documenti dell'archivio di stato di Mantova," *Studi Colombiani* (Genoa: SAGA, 1952), 2: 489-90.

Cites letters and other late 15th-c and early 16th-c documents referring to the exploits of CC, Vespucci, Magellan and Pigafetta, Cortés, Pizarro, etc.

[85] **Demetrio Ramos Pérez.** "Sobre la 'relación' de Pané dedicada a los taínos y su utilización por Martír de Anglería en 1497," *Archivo Hispalense* (Seville), 68, nos. 207-208 (1985): 419-29.

Sketches some of the problems, such as the date of Pané's account; the tardy delivery of the account to CC by Peralonso Niño; and Peter Martyr's access to the account prior to his use of it.

[86] **Aldo Agosto.** "La duplice redazione di un documento colombiano," *La Storia dei Genovesi: Atti del Convegno di Studi sui Ceti* (Genoa: Univ. degli Studi, 1987), pp. 133-39.

The act of 11 Oct 1496, in Genoa, which declares that Giovanni Colombo of Quinto will visit his cousin CC in Spain and that his 2 brothers, Matteo and Amighetto, will share his expenses, exists in a second "scratch" copy found in a bundle of papers in the Genoese archives by P. Repelto. This scratch copy was obviously written by the lawyer for his own use.

Rina Ferrarelli Provost

B. STUDIES OF DOCUMENTS OF COLUMBUS'S WRITINGS

1. General

[87] **Henry Harrisse.** "Autographs de Christophe Colomb récemment découvertes," *Revue Historique* (Paris), 51 (1893): 44-64.

Reviews and describes thoroughly the 16 docs specifically by or about CC in *Autógrafos de Cristóbal Colón* (1892), #6.

2. Studies of Documents of Individual Works by CC

[88] **Herbert Putnam.** "A Columbus Codex," *The Critic* (New Rochelle NY), 42 (1903): 244-51.

Description of the copy of CC's *Book of Privileges* (MS) that CC deposited with Fr. Gaspar Gorricio in the Las Cuevas Monastery in Seville. This is the Harrisse "Boston copy" (see #30). The "Boston" copy was later purchased from William Everett by the LC [where it is now held. It is now known as the "Washington copy"].

Cf. F.G. Davenport, #89, C.L. Nichols, #92, and C. Pérez Bustamente, #40.

[89] **Francis Gardiner Davenport.** "Texts of Columbus's *Privileges*," *Amer Hist Rev*, 14 (1909): 764-76.

Discusses interrelations of the codexes of CC's *Book of Privileges*. Demonstrates that the Veragua codex (not known to Stevens, #30) is the earliest extant form of the *Book*, largely compiled in 1497, before the 3rd voyage. The compilation of the *Book* had previously been assigned to 1502.

Cf. H. Putnam, #88, C.L. Nichols, #92, and C. Pérez Bustamente, #40.

[90] **Anonymous.** "Two Important Gifts to the New York Public Library by Mr. George F. Baker, Jr.: Columbus's Letter on the Discovery of America (1493-97) and Daniel Denton's Description of New York in 1670," *Bull N Y Pub Lib*, 28 (1924): 595-606.

"Important for Mr. [Wilberforce] Eames' list of the 17 15th-century editions of the Columbus letter, in Spanish, Latin, German, and Italian verse, with notes on existing copies and locations."

Mug 68

[91] **Angel de Altolaguirre.** "Autenticidad de la escritura de mayorazgo en la que don Cristóbal Colón declaró haber nacido en Genova," *Atti del XXII congresso internazionale degli Americanisti* (Rome) 2 (1926): 593-605.

The Spanish royal cedula of 28 Sept 1501 furnishes additional proof of the authenticity of CC's Mayorazgo of 22 Feb 1498, a document already accepted as authentic by the most eminent Spanish and foreign scholars of the 19th and 20th centuries.

[92] **C. L. Nichols.** "The Various Forms of the Columbus Codex," *Proc Mass Hist Soc* (Worcester MA), 59 (1926): 148-55.

Reviews the history of the four complete copies of CC's *Book of Privileges:* the two Genoa codexes, one now (1926) in Paris, one in Genoa; the Las Cuevas codex, in the Library of Congress; and the Veragua codex, presumably in the collection of the Duke of Veragua at the time of the article. A fifth, abbreviated copy of the *BoP* is known as the Providence codex, owned by the JC Brown Library.

Cf. H. Putnam, #88, F.G. Davenport, #89, and C. Pérez Bustamente, #40.

[93] **Cecil Jane.** "The Letter of Columbus Announcing the Success of his First Voyage," *Hisp Amer Hist Rev,* 10 (1930), 33-50.

Reviews the evidence concerning the identity of the person or persons to whom CC sent copies of the letter. Argues that CC sent the letter to the sovereigns through some unknown but appropriate intermediary, and that the sovereigns, wishing to publicize their claim to the new lands, endorsed copies of the letter to Santangel and Sánchez, who proceeded to publicize it in Spain and Italy respectively.

[94] **Randolph G. Adams.** *The Case of the Columbus Letter.* New York: Washington Square Book Club, 1939. 32 pp.

A study prepared as an address to the club. Examines the earliest printed editions of CC's "Letter" of March 4, 1493 and various forgeries of these editions.

C. E. Nowell

[95] **Samuel Eliot Morison.** "Texts and Translations of the Journal of Columbus's First Voyage," *Hisp Amer Hist Rev,* 19 (1939), 235-61.

Reviews (pp. 235-40) what is known of the original MS; judges the Las Casas abstract "well and honestly made" (239) and cites this abstract as the oldest and best text. Judges the 1892 De Lollis transcription much better than Navarrete's (1825). Judges the 1828 French translation by Verneuil and de la Roquette untrustworthy. The 1827 English translation by Samuel Kettell is also untrustworthy, but the corrected 1931 edition (by Lawrence and Young) is one of the best. The Markham-Bourne translation (1898, 1906) is careless and inaccurate, and inadequately annotated. Thacher's 1903 translation is also weak and inaccurate. Cecil Jane's 1930 translation is reasonably accurate except in dealing with the many technical matters of sailing, and of fauna and flora. Concludes that a new translation (his own) is justified.

[96] **Antonio Muro Orejón.** "Cristóbal Colón: el original de la capitulación de 1492 y sus copias contemporaneas," *Anuario de Estudios Americanos* (Seville), 7 (1950): 505-515.

An account of the copies made of the *Capitulations of Santa Fé* of 17 Apr 1492 from the now-lost original possessed at that time by CC. Prints the copy of 1497 made in Burgos at the time of the confirmation of CC's privileges. Includes notes recording textual variants.

[97] **Angela Codazzi.** "Di una versione italiana manoscritta della lettera di Cristoforo Colombo al 'Thesorero de su Mta.'" *Studi Colombiani*, 2 (Genoa: SAGA, 1952), 469-478.

Describes and analyzes an Italian version of CC's "Letter" of March 4, 1493, held in the Biblioteca Nazionale Centrale V. Emanuele in Rome. The letter, on paper of the period 1492-97, is closer to the extant Spanish versions than to the other versions in Italian or Latin. Its text is full of corruptions, perhaps deriving at least in part from being copied from an MS, and in part from the limited culture of the writer, probably a Lombard. The letter purports to deal with "*four* caravels of the King of Spain."

[98] **Emiliano Jos.** "El Diario de Colón: su fondamental autenticidad," *Studi Colombiani* (Genoa: SAGA, 1952), 2: 77-99.

Attempts to cut through the misleading aspects of the early documents—including especially those of Ferdinand C's *Historie*—to establish the importance and reliability of CC's *Journal* as an historical document.

[99] **Rafaello Morghen.** "Il Libro d'Ore di Cristoforo Colombo," *Studi Colombiani* (Genoa: SAGA, 1952), 2: 295-99.

Develops the evidence that a codex in the Biblioteca Corsiniana in Rome, which purports to be a book of devotions presented to CC by Pope Alexander VI, is genuine and not a hoax as argued by Cesare de Lollis and others.

[100] **Carlos Sanz.** *El gran secreto de la carta de Colón (crítica histórica) y otras adiciones a la Bibliotheca Americana Vetustissima.* Madrid: Librería General, 1959. 523 pp.

El gran secreto calls CC to task for self-aggrandizement in revealing his discovery to the Portuguese and in contriving through the *Letter to Santangel and Sánchez* (4 Mar 1493) to popularize himself throughout the world, acts that finessed the monarchs' privilege of deciding whether to keep the discovery secret. Sanz strongly applauds M. A. Pinzón for writing to the monarchs from Bayona to request an audience, instead of grandstanding like CC.

Otras adiciones adds a number of items, mostly copies and reprints of CC's *Letter* of 4 Mar 1493, which eluded Harrisse in his *BAV*, #731.

[101] **B. García Martínez.** "Ojeada a las Capitulaciones para la conquista de América," *Rev Hist Amer* (Mexico City), no. 69 (1970), pp. 1-40.

A speculation on the relationship between the terms conceded to CC and the evolution of the Spanish conquest and occupation of the New World. Studies 73 other agreements between the crown and the explorer-conquerors following CC, and details 105 clauses defining the agreed-on activities and limitations of these agents of the crown. An imposing study.

[102] **Antonio Rumeu de Armas.** "El *Diario de a Bordo* de Cristóbal Colón: el problema de lal paternidad del extracto," *Rev Indias* (Madrid), 36 (1976): 7-17.

Suggests that the strong differences among the *Journal,* Ferdinand C's *Historie,* and Las Casas' *Historia,* especially in recounting the last days before the landfall (and the struggle at that time between CC & crew) make it impossible that the *Journal* was the work of Las Casas.

[103] **Robert H. Fuson.** "*The Diario de Colón:* A Legacy of Poor Transcription, Translation, and Interpretation," *Terr Incog,* 15 (1983): 51-75.

Rpt. *In the Wake of Columbus* (Detroit: Wayne State Univ. Press, 1985), pp. 51-75. Begins with a succinct history of the *Journal* and of Las Casas' abstract; traces the transcriptions, translations, and editions of the abstract from Navarette through Oliver Dunn and Eugene Lyon; discusses at length transcriptive, translative, and interpretive flaws in these publications. "Although some errors were in the holograph original, and others appeared in the abstract, the greatest source of error is to be found in the collective work of those who have transcribed and translated the Las Casas narrative. It is the closest thing we have to a primary source, and should take precedence over *any* later commentary . . . " (p. 74).

[104] **Juan Pérez de Tudela.** *Tratado de Tordesillas. Study.* Madrid: Testimonio Compañia, 1985. 96 pp.

A study of the text of the 1494 treaty between Spain and Portugal. Concludes that the treaty achieved the intended effects (1) of protecting the Atlantic flank of the Portuguese exploitation of the African route to the Indies, and (2) of allowing Spain to proceed without interference in exploiting the newly discovered lands across the Atlantic. Appends a description and transcription of the treaty by T. María Martínez and J. M. Ruíz Asencio, pp. 51-64. English translation of Pérez de Tudela's essay, pp. 67-96.

[105] **Antonio Rumeu de Armas.** *Nueva luz sobre las Capitulaciones de Santa Fé de 1492 concertadas entre los Reyes Católicos y Cristóbal Colón: estudio institucional y diplomático.* Madrid: Consejo Superior de Investigaciones Científicas, 1985. 276 pp.

A study of the versions and the registry of the treaty of 17 Apr 1492 between the Spanish monarchs and CC. Concludes that the document was registered only in Aragón, and only after CC's return in 1493. Suggests that material was interpolated into the document after CC's return. Rejects all traditional associations of the document with the court of Castile, and thus casts doubt on the validity of many traditional assertions about the *Capitulations.*

[106] **Demetrio Ramos Pérez.** *La primera noticia de América.* Valladolid: Casa-Museo de Colón, 1986. 147 pp.

Cuadernos Colombinos, 14. A study of the various versions of CC's *Letter* (To Santangel and Sánchez) of 4 Mar 1493. Ramos concludes (p. 114) that the curious differences in the extant versions reflect Iberian adaptation to King Ferdinand's astute politics.

[107] **Delno C. West.** "Wallowing in a Theological Stupor or a Steadfast and Consuming Faith: Scholarly Encounters with Columbus' Libro de las Profecías," *Columbus and his World* (Ft. Lauderdale, FL: CCFL, 1987), pp. 45-56.

The *Libro de las profecías* was drafted in 1501 and remains the only work written by CC which has not been thoroughly studied. This paper describes the document and argues that the persuasive rhetorical power of noted 19th and 20th-century historians convinced scholars that the treatise was the ravings of a troubled mind. Only recently have historians begun to understand the importance of this work as it relates to Columbus's mentality and Enterprise of the Indies.

DCW

[108] **Carlos Sanz.** *En realidad cuando se descubrió América?* No place: Carlos Sanz, no date. 7 pp.

Study of CC's letter of 15 Feb – 4 Mar 1493 announcing the discovery. Describes 17 editions published in various European countries.

C. STUDIES OF OTHER DOCUMENTS NOT BY COLUMBUS HIMSELF

[109] **Giuseppe Pennesi.** "Pietro Martire d'Anghiera e le sue relazioni sulle scoperte oceaniche," *Raccolta*, V.2 (1894): 7-10.

A brief life and an account of P. Martyr's narratives respecting CC's voyages, in his *Decades* and other writings.

[110] **George E. Nunn.** "The *Imago Mundi* and Columbus," *Amer Hist Rev*, 40 (1935), 646-61.

Studies the postils in CC's copy of Pierre d'Ailly's *Imago Mundi*, with reference to Buron's edition and the transcription in the *Raccolta*. Points out that, despite Vignaud, the postils correct or augment the text in 80 cases. CC's understanding of the terrestrial degree and his geographical conception of eastern Asia differ radically from d'Ailly's, but on the 3rd voyage CC did act on d'Ailly's assertion of the nearness of Africa to India by sea by starting at the Cape Verdes in order to strike Marco Polo's great island 750 miles southeast of Asia. See Cornelio Desimone, "Quistioni [sic] Colombiane," #760, p. 45.

Mug 77

[111] **Leonardo Olschki.** "Hernan Pérez de Oliva's 'Ystoria de Colón,'" *Hisp Amer Hist Rev*, 23 (1943): 165-96.

Written in conjunction with Olschki's 1942 transcription of the unique MS in the possession of Frank Altschul of New York. Describes the 34-sheet, late 16th-c MS of the first life of CC in Spanish (prior to 1531). Abstracts the entire work (174-82), which loses its biographical character as it proceeds and becomes "an impersonal account of strange, pathetic, and extraordinary events" (p. 174) on the first 3 voyages, and follows with a final chapter on the religion, rites, and superstitions of the West Indians. Olschki observes that Pérez de Oliva is not influenced by Ferdinand Columbus's *Historie*, but heavily by Peter Martyr's *Decades*, and does not introduce any material not already available elsewhere. Nonetheless the narrative shows a keen and lyrical sense of the conflicts experienced by the helpless American natives. Olschki concludes that F. Columbus's *Historie* may have been written to refute the "false" emphases of just such lives as this one.

[112] **Alfonso García Gallo.** *Los origines de la administración territorial de las Indias.* Madrid: Instituto Francisco de Vitoría, 1944. 99 pp.

A study (1) of the *Capitulations of Santa Fé* of 30 April 1493; (2) of the administrative offices thereby endowed on CC (admiral, vice-regent, governor); (3) of the powers and scope implied in these offices; (4) of the hereditary character of these offices; (5) of the lack of supporting offices provided for CC, whose rule was personal and absolute; and (6) of the delegated territorial offices established by CC (lieutenant, adelante, alcalde mayor) to administer his official powers in his absence.

[113] **Henry R. Wagner.** "Peter Martyr and his Works," *Proc Amer Antiq Soc* (Worcester MA), n.s. 56 (1946): 239-88.

"Martyr's published writings, pp. 277-78."

"Much of value concerning the first installment of *De orbe novo decades*, pub. by Martyr in his *Opera*, Seville, 1511, although W. is principally interested in Martyr as a source for Cortés's conquest of Mexico."

Mug 101

[114] **Antonio Muro Orejón.** "Cristóbal Colón: el original de la Capitulación de 1492 y sus copias contemporaneas," *Anuario Estud Amer* (Seville), 7 (1950): 505-15.

History of the original of the *Capitulations of Santa Fé* of 17 Apr 1492, the basis of CC's claims to office, rank, share in profits (and the hereditary status of these) upon successfully claiming lands for the Catholic monarchs in the western ocean. Account of early copies. Photos of 5 copies still extant, and transcription of the copy inserted in the *Privilegio real* of 1497. Account of variants in the copies.

[115] **Gastone Imbrighi.** "Sui rapporti tra le 'Historie' attribuite a D. Fernando Colombo e la 'Historia' scritta dal vescovo Bartolomé de las Casas," *Studi Colombiani* (Genoa: SAGA, 1952), 2: 567-79.

A comparison of the *Historie* and the *Historia* leads to these conclusions: (1) the *Historie* was prepared not to exalt and glorify CC's achievements but simply in resentment against those who belittled CC's achievements and wished to diminish the privileges of CC's heirs; (2) the *Historie* frequently derives from and summarizes Las Casas' *Historia,* while Las Casas does not use the *Historie* as a source; (3) any discrepancies or contradictions between the two histories are due to the "pseudo-Ferdinand" who prepared Ferdinand's incomplete work for publication, probably the unprincipled Luis Colón, CC's grandson.

[116] **Assunto Mori.** "Giuseppe Pagni e il suo manoscritto inedito su Cristoforo Colombo e Amerigo Vespucci," *Studi Colombiani* (Genoa: SAGA, 1952), 2: 491-4.

Describes a collection of documents significant for the restoration of V's reputation *vis-á-vis* CC.

[117] **Alejandro Cioranescu.** *Primera biografía de Colón: Fernando Colón y Bartolomé de las Casas.* Tenerife: Aula de Cultura, 1960. 252 pp.

Argues that Ferdinand C did not write much of the biography entitled the *Historie* (Venice, 1571), but that the book consists mostly of excerpts from an early draft of Las Casas' *Historia de las Indias.* Attributes the fraud to CC's grandson Luis. Cf. #115.

Martin Torodash

IV

Columbus's Life

A. GENERAL TREATMENTS

1. Biographies

[118] **William Robertson.** *The History of America.* London: Strahan, Cadell, 1777. 2 vols.

Bk. 2, I.59-175, a ground-breaking, straightforward account of CC's life and voyages, based on Oviedo, Peter Martyr, Gómara, Herrera, and Bernáldez. Menéndez Pelayo (#759) says it served as the basis of all the popular biographies of CC that appeared in profusion at the end of the 18th c., i.e., at the time of the 3rd centennial.

Bk. 1, I.1-57, Classical and medieval background; 38-57, the Portuguese activity in the Atlantic to the return of B. Diaz from the Cape of Good Hope. Bk. 3, I.177-245, Spanish exploration and settlement, 1504-1518, to the Cortés expedition.

[119] **William Russell.** *The History of America, from its Discovery by Columbus to the Conclusion of the Late War.* London: Fielding and Walker, 1778.

Ch. 2, a compact account of what could be known of CC's life and voyages and of the voyages of Spaniards to the time of CC's death. Repeated references to Ferdinand's *Historie* and Herrera, and also to Oviedo, Peter Martyr, and William Robertson, #118. Ch. 1, the maritime background to 1486, when the Portuguese dominated maritime affairs in the Atlantic.

[120] **Jeremy Belknap.** *A Discourse intended to Commemorate the Discovery of America by Christopher Columbus.* Boston: Belknap and Hall, 1792.

A sketchy account of CC's life and discoveries, citing mostly Ferdinand C's *Historie*, but also Herrera and Peter Martyr. Probably also indebted to William Russell, #119, and/or Robertson, #118.

Concludes with an appeal for a more Christian treatment of American Indians in the newly founded USA.

[121] **Juan Bautista Muñoz.** *Historia del nuevo-mundo.* Madrid: Viuda de Ibarra, 1793. Vol. 1 (only vol. published). 263 pp.

To 1500. After a brief background describing the beginning of European ocean navigation and population of the Atlantic islands, relates CC's career from the inception of his enterprise in Portugal to the Roldán rebellion. Account based on the documents made available to him in the Spanish archives by the Spanish king, and on those located by his own diligent research in the archives of Simancas, Seville, Cádiz, Madrid, the Escorial, the monastery of Montserrat, the universities of Salamanca and Valladolid and various cathedrals and convents in Spain, and the Torre do Tombo in Lisbon. Menéndez Pelayo calls this biography "la biografía más clásica y mejor escrita que en castellana tenemos del Almirante" (1892), #759, p. 103.

[122] **Washington Irving.** *A History of the Life and Voyages of Christopher Columbus.* New York: Carvill, 1828. 4 vols.

1st British edition, London: John Murray, 1828. A landmark biography based on #2 and on other Spanish documents, especially those in the archives of the then-current Duke of Veragua. The strengths of the work are typified by the note (I,96) clarifying the confusing testimony of García Fernández in the *Pleitos,* by suggesting that Fernández conflated two of CC's visits to La Rábida, making them seem to be a single visit. The weaknesses are typified by Irving's fanciful and sentimental account (I,117-31) of how CC's laudably enlightened presentations to Talavera's royal commission at Salamanca were rejected because of benighted religious prejudice.

[123] **Alexander Freiherr von Humboldt.** *Examen critique de l'histoire et de la géographie du nouveau continent et des progrès de l'astronomie nautique aux quinzème et seizième siècles.* Paris: Gide, 1836-39. 5 vols.

Contains classic account of CC's career. Chapters on CC's goals in his voyages of discovery; evolution of cosmographical ideas prior to CC; CC's cosmographical ideas, and the motives of his Enterprise; ancient cosmology and cartography; influence of Toscanelli on CC's projects; Martin Behaim and Magellan; CC and Behaim; first discoveries on the east coast of America; influence of the shape of Africa on ideas about the shape of America; secret expeditions; motives for the discovery of America at the end of the 15th c.; considerations on the physical geography of the earth and on pre-Columbian contacts with America; Scandinavian voyages of the 12th and 13th c.; CC unaware of the Sc. voyages; social state of America prior to CC; various pre-Columbian voyages;

cosmography in the Middle Ages; St. Brendan's isle; Antillia and the Isle of the 7 Cities; the isle of Brazil, and other minor matters; probable communication between the 2 worlds because of air and ocean currents.

[124] **Aaron Goodrich.** *A History of the Life and Achievements of the So-Called Christopher Columbus.* New York: Appleton, 1874. 403 pp.

A notable piece of Columbus-bashing, balancing at the other extreme Roselly de Lorgues' attempt, #655, to make CC a saint. Goodrich does a thorough job of ferreting out CC's unattractive qualities, to the exclusion of much of his greatness.

Nonetheless, the book is frequently judicious and thoughtful, as in its analysis of Vespucci and his friendly relationship with CC. Hardly any writer of the day had read more on the subject of CC, and few managed to mention more issues.

[125] **Henry Harrisse.** *Christophe Colomb: son origine, sa vie, ses voyages, sa famille & ses descendants, d'après des documents inèdits tirés des archives de Génes, de Savone, de Séville et de Madrid.* Paris: E. Leroux, 1884. 2 vols.

A model scholarly biography, except that the book is heavily tinged with Harrisse's distrust of Las Casas. Introduction reviews (a) source documents, Mss & printed, through Muñoz and Navarrete; (b) source histories from Giustiniani and other contemporary Genoese to Luigi Bossi and Washington Irving. The chapters: origins of CC's family; the family itself; CC's life to the first voyage; his 4 voyages; his last years and death; his siblings; his son, daughter-in-law, and grandson; CC's legitimate lines of descent; Ferdinand Columbus; the various illegitimate lines stemming from Diego's and Luis's natural children; and supposed relatives. Supplies genealogical charts.

Appendices relate many things that do not fit exactly into the chapters themselves.

[126] **José M. Asensio Toledo.** *Cristóbal Colón: su vida, sus viajes, sus descubrimientos.* Barcelona: Espasa y Compañía, n.d. [1888]. 2 vols.

A very elaborate biography, in an edition splendid indeed. The scholarship, however, was vigorously attacked in the review by Henry Harrisse, "Christophe Colomb et ses historiens espagnols," *Rev Critique de l'Historie et de Litterature* (Paris), 26 Sept – 3 Oct 1892; rpt. Paris, 1892.

[127] **Justin Winsor.** *Christopher Columbus and How he Received and Imparted the Spirit of Discovery.* Boston and New York: Houghton-Mifflin, 1891. 674 pp.

A full biography based on the documents available prior to the *Raccolta*, #8, with judicious appraisal of prior critical study. At the end, chapters on CC's descendants (pp. 513-28) and on the results for geography of CC's life (pp. 529-60). P. 111, reprints O. Peschel's reconstruction of the Toscanelli map.

[128] **John Fiske.** *The Discovery of America with some Account of Ancient America and the Spanish Conquest.* Boston and New York: Houghton Mifflin, 1892. 2 vols.

An enormously detailed and learned attempt to place CC's life and voyages in the context both of European history and of what could be known about aboriginal America at the time.

Chapters: Ancient America; Pre-Columbian Voyages; Europe and Cathay; The Search for the Indies; The Finding of Strange Coasts; Mundus Novus (A compendious report on post-1493 European explorations resulting from the discovery); The Conquest of Mexico; Ancient Peru; The Conquest of Peru; Las Casas; The Work of Two Centuries.

Special emphasis on the evolution of the concept of the "New World"; cf. Juan Valera, "Concepción progresivo del nuovo mondo," #489. Fiske reproduces Peschel's 1867 reconstruction of Toscanelli's map, I,356.

The analytical table of contents is a model of detailed outline. This book, limited only by the author's lack of access in 1891 to the *Raccolta*, #8, and to Vignaud's *Études Critiques*, #142, and *Histoire Critique*, #433, is among the very best treatments of the subject.

[129] **Filson Young.** *Christopher Columbus and the New World of his Discovery,* 3rd ed. New York: Henry Holt, 1912.

A thorough revision of the earlier editions (1st ed. 1906). Prefaced by an approving letter from Henry Vignaud. A full and careful biography based on the available documents, with a series of useful appendices. See especially the long note on navigation by the Earl of Dunraven, pp. 339-422, #659.

[130] **H. H. Houben.** *Christopher Columbus: The Tragedy of a Discoverer.* tr. from the German by John Linton. London: George Routledge, 1935. 308 pp.

A quasi-fictional account following the main lines of authentic history, and pursuing many poignant ironies of CC's career. The translator's style probably influenced Morison's *AOS*, #132, as in the chapter title "The Hell on Hispaniola" (Morison, "Hell in Hispaniola"). Far superior to most fictional treatments of CC's life.

[131] **Daniel Sargent.** *Christopher Columbus.* Milwaukee: Bruce, 1941. 214 pp.

Analysis and depiction of CC's life and times. Notable for representation of Genoa, Portugal, and Spain of the late 15th century.

Martin Torodash

[132] **Samuel Eliot Morison.** *Admiral of the Ocean Sea: A Life of Christopher Columbus.* Boston: Little, Brown, 1942. 2 vols.

A lucid, highly readable account, with special emphasis on the voyages and special attention to navigational matters. The notes contain a wealth of useful bibliographical references. In the interest of a lively narrative, there is some blurring of the line between fiction and documented fact, as in the chapter "In Castile," I,107-29.

The Pulitzer Prize-winning abridgement, same title (Boston: Little Brown, 1942), 680 pp, omits (1) the preliminary chapter on ships and sailing; (2) all the scholarly notes; (3) the chapter on syphilis, II, 193-218, #603. Each chapter shortened, and index reduced from 17 to 7 pages.

[133] **Enrique de Gandía.** *Historia de Cristóbal Colón. Análisis crítico.* Buenos Aires: Claridad, 1942.

"A workmanlike, common-sense examination of Columbian problems, but little beyond First Voyage." Morison, *AOS*, II,22.

[134] **Antonio Ballesteros Beretta.** *Cristóbal Colón y el Descubrimiento de América.* Barcelona and Buenos Aires: Salvat, 1945. 2 vols.

In *Historia de América y de los pueblos Americanos*, Vols. 4 & 5. By far the fullest and most comprehensive life of CC, based on the primary documents and on comparative study of the secondary literature. Bibliographies full but difficult to use. Maps borrowed from Morison's *AOS*. No index.

[135] **Bjorn Landström.** *Columbus: Historien om Amiralen över Oceanen Don Cristóbal Colón.* Stockholm: Bokförlaget Forum, 1966. 208 pp.

English trans., *The Story of Don Cristóbal Colón, Admiral of the Ocean Sea, and his Four Voyages Westward to the Indies, According to Contemporary Sources.* New York: Macmillan, 1967. 206 pp.

A profusely and handsomely illustrated life that appraises CC's abilities, theories, and actions. Follows Morison's conclusions about the standard issues of birth, provenance, landfall, etc., but gives CC more credit for using instruments to navigate. Also credits CC with guessing the Atlantic wind-pattern and with more missionary zeal than greed for wealth and power. Excellent drawings, especially those representing ships.

[136] **Cesare De Lollis.** *Cristoforo Colombo nella leggenda e nella storia.* Florence: Sansoni, 1969.

A re-publication of the famous 1892 account of CC's life by the distinguished textual editor of the 1892-96 *Raccolta.* Roberto Almagià's 1931 preface is reprinted here, pp. vi-xxii, followed (pp. xxiii-xxvi) by an updating of the critical milieu by Elio Migliorini. Within the context of the prefaces, this book has a place among the seminal biographies of CC.

The appendix includes six of De Lollis's critical essays on CC.

[137] **Jacques Heers.** *Christophe Colomb.* Paris: Hachette, 1981. 666pp.

Opens new vistas in CC biography, first with wide-ranging application of 15th-c history, including Genoese activity not only at home but in Spain, Portugal, Madeira, and the Canaries; and second, with an analytical rather than chronological approach to biography. Cf. #383.

2. Studies of Matters that Concern
More than One Period of Columbus's Life

[138] **Henry Harrisse.** *Christopher Columbus and the Bank of St. George.* New York: Privately printed, 1888. 126 pp.

Italian ed.: *CC e il Banco di S. Giorgio* (Genoa: Municipio, 1890). An essay on CC's happy relationship with the Ufficio di San Giorgio in Genoa, which also discusses the bank's operations in medieval times and the documentary evidence that CC was Genoese.

[139] **Césareo Fernández Duro.** "Investigación de los bienes de fortuna que tuvo Cristóbal Colón," *El Centenario* (Madrid) 1 (1892): 69-83.

A classic statement (frequently ignored) of the various evidence demonstrating that CC not only did not die in poverty, but continued to receive remarkable support from the crown even after being returned from Hispaniola in irons in 1500, and died a relatively wealthy man. Cf. Juan Gil's article #169.

[140] **Henri Harrisse.** *Christophe Colomb devant l'Histoire.* Paris: Welter, 1892. 124 pp.

Reviews current truths and falsehoods about CC, and fixes his greatness in his genius for audacious pursuit of a great dream in such a way as to direct the subsequent evolution of world history and culture.

[141] **G. Marcel.** "Christophe Colomb et Beatrix Enriquez de Arana," *La Géographie* (Paris), 5 (1902): 376-78.

Reviews several documents found in the Colombina library in Seville and the notarial archives of Córdova which demonstrate that Beatriz and her family were of very humble extraction and that she was poor when CC met her and poor when he died, and that she died still unmarried. Debunks the frequent attempts, at the time of the 4th centennial, to show that CC married her, that she was of a noble family, and that it was she who paved CC's road to favor with the queen.

[142] **Henry Vignaud.** *Études critiques sur la vie de Colomb avant ses découvertes.* Paris: Welter, 1905. 543 pp.

A patient, scholarly assemblage of facts. Topics studied: The origins of CC's family; the two Colombos (Guillaume de Casenove, dit Coullon; and George the Greek, also known as Bissipart and Colombo Junior); the true date of CC's birth (1451); education and early adventures; arrival in Portugal and the naval battle of Aug 1476; CC's voyage to the Arctic; his residence in Portugal; his marriage and his Portuguese family. Voluminous documentation.

[143] **Henry P. Biggar.** "The New Columbus," *Annual Report of the American Historical Association for 1912* (Washington DC: AHA, 1914), pp. 95-104.

Attacks Henry Vignaud's perennial thesis that CC's objective was not Asia, citing (a) his proposal to John II of Portugal; (b) the evidence of João de Barros; (c) CC's passport letter, apparently for the Grand Khan; (d) the introduction to the *Journal* of the first voyage.

Mug 7

[144] **Rudolf Cronau.** *The Discovery of America and the Landfall of Columbus. The Last Resting Place of Columbus.* New York: R. Cronau, 1921. 89pp.

Landfall at Watlings Island, burial in Santo Domingo cathedral. Maps, illustrations, plates.

Mug 17

[145] **Henry Vignaud.** *Le vrai Christophe Colomb et la légende.* Paris: Picard, 1921. 230 pp.

Restates Vignaud's view, developed through a lifetime of research, that CC did not set out to find the East Indies but only to find islands in the Atlantic.

John Bigelow

[146] **José de la Torre y del Cerro.** *Beatriz Enríquez de Harana y Cristóbal Colón.* Madrid: Compañia Iberoamericana de Publicaciones, 1933. 181 pp.

Publishes new documents about the family of Beatriz, the mother of CC's second son Ferdinand. Infers that CC's affair with her would not have been seen as discreditable, and that Beatriz herself was sufficiently loose to lose the affection of CC and Ferdinand too.

[147] **P. Gribaudi.** "Il padre Gaspare Gorricio di Novara, amico e confidante di Cristoforo Colombo," *Bolletino della Reale Diputazione Subalpina di Storia Patria* (Torino), 17 (1938): 1-87.

An account of Fr. Gorricio's life, along with information on his brothers Francesco and Melchiore, emigrants from Novara, north of Genoa, to Spain. Chapters treat: (1) The Gorricio brothers in Spain; (2) The family in Novara; (3) Fr. G. and CC during the 3rd voyage; (4) Fr.G and CC between voyages 3 & 4; (5) Fr. G. & CC during the 4th voyage; (6) *The Book of Prophecies* & Fr. G.; (7) Fr. G & CC's brothers; (8) Fr. G. & Diego C's first testament; (9) Fr. G and the Columbus Archive of Las Cuevas; (10) Final information about Fr. G.

[148] **Manuel Giménez Fernández.** "Nuevas consideraciones sobre la historia y el sentido de las letras alejandrinas de 1493 referentes a las Indias," *Anuario Estud Amer* (Seville), 1 (1944): 171-427.

Elaborately documented and illustrated study proceeding from suggestions in the documents that the church-state conflict in Hispaniola between CC and Fray Bernal Buil actually derived from a secret agreement between King Ferdinand and Buil to limit the powers awarded to CC in the *Capitulations of Santa Fé.* Giménez infers that the bulls were sought by Ferdinand (a) as a practical expedient to co-opt or pre-empt Portuguese claims to the area discovered by CC and (b) to place limitations on CC's power in the new lands. Later, the bulls served historians in defending Spanish actions as well as Las Casas in attacking them.

[149] **Francis Borgia Steck.** "Christopher Columbus and the Franciscans," *The Americas: A Quarterly Review of Inter-American Cultural History* (Wash. DC), 3 (1946-47): 319-41.

Reviews chronologically CC's contacts, definite and possible, with the Franciscans from the time of his arrival in Portugal in 1476 to his death. Treats the Portuguese Franciscans, the Spanish Franciscans before CC's embarcation from Palos in Voyage 1, the Franciscans' part in the 4 voyages, and CC's relationship with the Franciscans in the final two years of his life.

[150] **Henry R. Wagner.** "Marco Polo's Narrative becomes Propaganda to Inspire Colón," *Imago Mundi* (Leiden), 6 (1949): 3-13.

CC's direct knowledge of Marco Polo in 1492 is not certain, but he had read Toscanelli's letter to Martins and may well have learned there of Cipangu and Cathay.

[151] **Armando Alvarez Pedroso.** "Recuerdos Colombinos de la Republica Dominicana," *Studi Colombiani* (Genoa: SAGA, 1952), 3: 31-44.

A description of and commentary on sites and buildings associated with CC in Santo Domingo (San Nicolás church, Dominican Convent church, monastery and church of San Francisco, Torre de Homenaje, San Miguel church, Alcázar Colón, and the cathedral). Others: ruins at the presumed site of Isabela; the hill "Santo Cerro" where CC and Bartholomew successfully withstood an attack by native rebels.

[152] **Francisco Domínguez Compañy.** "Colón y los Indios," *Studi Colombiani* (Genoa: SAGA, 1952), 2: 691-96.

CC's initial justification for taking and selling Indians as slaves included these points: (1) that slavery would be only a temporary measure; (2) its purpose would be educational, teaching the slaves Spanish and thus expediting their conversion to Christ; (3) it would apply only to the cruel and incorrigible cannibals; and (4) it would be used as the only available way to pay the expenses of the colony. All but the last of these qualifications soon disappeared from CC's policy. To the credit of the crown, the monarchs never accepted these arguments, but held that the only justifiable slavery was of captives in a just war.

[153] **Pedro de Leturia.** "Ideales político-religiosos de Colón en su carta institucional de 'Mayorazgo': 1498," *Studi Colombiani* (Genoa: SAGA, 1952), 2: 249-72.

The Mayorazgo develops three politico-religious ideals: (1) the reconquest of Jerusalem; (2) faith in the Pope; (3) the formation of sacred cultural and charitable institutions in Santo Domingo.

[154] **Anibal Ybarra Rojas.** *Colón y los americanos.* León de Centroamérica: Talleres Nacionales de Nicaragua, 1952. 996 pp.

An extremely detailed study of the life of CC from the point of view of the native Americans. Anticipates by at least a decade the current intense concern with the confrontation between Europeans and American aborigines.

[155] **Pietro Scotti.** "Concetti etnologici di Colombo," *Studi Colombiani* (Genoa: SAGA, 1952), 2: 119-25.

Reviews the ethnological assertions about the New World in CC's *Journal* and other writings; praises the *Journal* for its clarity and precision; and attributes to CC the founding of American ethnology. Cites CC's participation in identifying the pre-Columbian cultures of the Siboney, Arawaks, and Caribs.

[156] **Louis-André Vigneras.** "New Light on the 1497 Cabot Voyage to America," *Hisp Amer Hist Rev,* 36 (1956): 503-509.

Reports discovering in the Archive of Simancas a letter of the year 1497 from the Englishman John Day to the Almirante Mayor of Castile, whose office was held at the time by Don Fadrique Enrique. The letter reports the John Cabot voyage of 1497, giving the latitudes of discovery, 45 deg. to 51 deg. 30 min., including Nova Scotia and Newfoundland. Day mentions that he is sending a map of the area discovered (perhaps, V. notes, a source used by Juan de la Cosa in his map of the world dated 1500). Day also mentions an unsuccessful voyage to the same area, perhaps in 1496. Finally, Day mentions an earlier discovery of the same land "in other times."

Vigneras reprints the letter (pp. 507-509), in which Day also reports that he is sending a copy of Marco Polo (see #157). Later speculation has suggested that this letter is really to CC; see David B. Quinn, "John Day and Columbus," *Geog Jour* (London), 133 (1967): 205-209, #159.

[157] **Louis-André Vigneras.** "The Cape Breton Landfall: 1494 or 1497. Note on a letter from John Day," *Canadian Hist Rev* (Toronto), 38 (1957): 219-29.

Includes a translation of the John Day's letter (published in Spanish in #156), perhaps to CC, in which Day mentions sending a copy of *Marco Polo's Travels*. V. suggests that this book may well be the one in Ferdinand's library, containing 366 marginal notes by CC. [This suggestion has led some contemporary scholars to postpone until the late 1490's the dating of CC's notes in his copy of Marco Polo. See Juan Gil, items #77 and 78.]

[158] **Antonio Rumeu de Armas.** "Cristóbal Colón y Beatriz de Bobadilla en las antevísperas del descubrimiento," *El Museo Canario* (Las Palmas), nos. 75-76 (1960), pp. 255-79.

Asserts that CC knew Beatriz's mother-in-law Inez Peraza in Andalusia in the early 1590's; that Beatriz, seeking to establish her son's claim to the island of Gomera and to quash charges against herself for cruelty to the natives of Gomera, was in Andalusia 5 times in the period 1590-92; that she and CC very likely knew each other in Santa Fé, Córdova, and/or Puerto de Santa María; and that since CC when he got to the Canaries in Aug 1492 knew of the ship that Beatriz had chartered from Anton de Grájeda (and hoped to substitute it for the damaged *Pinta*), it is very likely that CC was in contact personally with Beatriz in Puerto de Santa María in the summer of 1492. Concludes in view of this that Michele da Cuneo was very likely right in asserting in his letter on the 2nd voyage that CC was in love with Beatriz.

[159] **David B. Quinn.** "John Day and Columbus," *Geog Jour* (London), 133 (1967): 205-209.

Discusses the strong probability that the John Day letter to the "Almirante" about the 1497 John Cabot voyage to America was indeed to CC, a part perhaps of a two-way correspondence wherein CC received and perhaps accommodated new ideas that might have affected his view of his discovery when he sailed on the 3rd voyage in 1498. Suggests that CC at this point may not yet have completely closed his mind to other possibilities besides the identity of his discoveries with the islands of the Far East. The Day letter was discovered by L.-A. Vigneras. See "New Light on the 1497 Cabot Voyage to America," *Hisp Amer Hist Rev*, 36 (1956): 507-509 (#156).

[160] **T. María Martínez.** "Cuestiones discutidas acerca de Colón expuestas y juzgadas por Las Casas," *Rev Indias* (Madrid), 29 (1969): 303-322.

Cites and quotes passages from Las Casas to register the Dominican's opinion on these prominent questions about CC: (1) CC's birthplace (Terra-rubia, in Liguria); (2) whether he was of Jewish extraction (no); (3) relations with the Pinzóns (CC was aided much by them, esp. by Martín Alonso, but came into extreme conflict with him over his behavior in the Indies, especially his capture of natives, whom CC set free. Rejects as unjust the allegations against CC in *Los Pleitos* alleging that the Pinzóns deserved the credit for the discovery); (4) the landfall (the light CC saw on the night of 11-12 Oct was on Guanahani, discovered several hours later; and CC was awarded the prize instead of Rodrigo de Triana because he saw the light before Triana saw land); (5) CC and Vespucci (attributes the discovery of S. America to CC; expresses vexation that it is called America, and accuses V of severe injustice to CC in claiming to have discovered it first himself); (6) CC and Francesco de Bobadilla (accuses B of injustice to CC in not advising CC of his full royal powers immediately on arrival in Hispaniola, an omission that led CC to claim powers that B's royal commission had cancelled; and sees B's action in imprisoning and chaining CC as a shocking insult to the viceroy); (7) CC's remains (after burial in Las Cuevas in Seville in 1509, they were by royal orders from Charles V exhumed and sent along with Diego's wife to Santo Domingo, and there interred in the cathedral near Diego's body).

[161] **Giuseppe Caraci.** "A proposito delle 'postille' colombiane," *Pubblicazione dell'Istituto di Scienze Geografiche* (Genoa), 18 (1971): 3-15.

CC's postils, which represent one of the few documentary direct Columbian sources of the conceptual genesis of the discovery of the New World, were not written while he was in Portugal, but during his sojourn in Spain, 1485-92. Attacks the contention of E. Buron in his 1930 edition of D'Ailly's *Imago Mundi* that note B 858c was written in 1481. Caraci is apparently unaware of David B. Quinn. "John Day and Columbus," *Geog Jour* (London), 133 (1967): 205-209, #159.

[162] **Cuadernos Colombinos.** Valladolid: Casa-Museo de Colón, 1971-86.

A series of monographs on topics closely related to CC. Authors and short titles of the volumes included in this guide: 1. Juan

Manzano, *Los mótines* (1971), #271; 2. Demetrio Ramos, *Los contactos transatlánticos* (1972), #415; 3. Demetrios Ramos, *Por que tuvó Colón* (1973), #440; 4. R. A. Laguardia Trias, *El enigma de las latitudes de Colón* (1974), #673; 5. J. Cortesão and A. Teixeira da Mota, *El viaje de Diogo de Teive; Colón y los Portugueses* (1975), #418 and #219; 6. L.-A. Vigneras, *La busqueda del paraíso* (1976), #416; 7. Demetrio Ramos, *Los Colón y sus pretensiones* (1977), #627; 9. Emiliano Jos, *El Plan y la génesis del desc.* (1980), #442; 10. Demetrio Ramos, *Las variaciones ideológicas* (1982), #351; 11. A. Milhou, *Colón y su mentalidad mesiánica* (1983), #637; 15. Luis Fernández Martín, *El Almirante Luis Colón* (1986), #631.

[163] **Joaquín Balaguer.** *Colón, precursor literario*, 2nd ed. Santo Domingo: Privately published, 1974. 150 pp.

3rd ed., 1986. 3 sections: (1) "Cristóbal Colón," pp. 9-41. Observations on CC's felicitous writing style and gift for poetic expression. (2) "El sentimiento de la naturaleza en Colón y los historiadores de las Indias," pp. 45-65. Compares CC's accounts of natural phenomena with those of Las Casas, Oviedo, Vespucci, and the poets of the conquest. (3) "Los dominicos en la Española," not about CC.

[164] **Ilaria Luzzana Caraci.** "La postilla colombiana B 858 e il suo significato cronologico," *Atti II Conv Internaz Stud Col* (Genoa: CIC, 1977), pp. 197-223.

Calls for a joint, multi-disciplinary review and reorientation of the scholarly approach to CC's postils, in order to open up their significance as a source of knowledge about the navigator. Illustrates with a close analysis of the famous *Coenta* postil to D'Ailly's *Imago Mundi*, which has been widely understood as having been written in 1481, as meaning that CC could and did write Spanish in 1481, and as meaning that he was Jewish. Caraci argues that none of these conclusions is possible, that CC must have copied the note verbatim and not composed it himself, and that the postil cannot have been made by CC as early as 1481. The matter of determining the date when CC began to write his postils must be placed on a firmer basis.

[165] **Lazzaro María de Bernardis.** "Le bolle Alessandrine, San Roberto Bellarmino, e la 'potestas indirecta in temporalibus,'" *Atti III Conv Internaz Stud Col* (Genoa: CIC, 1979), pp. 547-64.

The 5 Alexandrine bulls of 1493 *in re* CC drew their authority from the 'potestas *directa*' inherent in the rigid application of the theocratic system.

[166] **Francesca Cantù.** "La presenza e l'azione di Cristoforo Colombo nel nuovo mondo nell'esperienza politica e nel giudizio storico di Bartholomé de las Casas," *Atti III Conv Internaz Stud Col* (Genoa: CIC, 1979), pp. 467-82.

Las Casas sees CC in a double light. He accepts him as God's choice to open the western hemisphere to European Christian culture; but he condemns CC's unchristian and rapine administrative policy toward the native Americans *in toto*, and concludes that CC adopts this policy for fear that the crown will abandon the enterprise if it does not show impressive profits. Las Casas thinks CC fell into this error through ignorance of divine and natural law, an ignorance not excusable even though CC was not an educated man.

[167] **Pietro Barozzi.** "Le postille colombiane al Milione," *Scritti geografici in onore de Aldo Sestino* (Florence: Soc. di Stud. Geog., 1982), pp. 53-65.

Two conclusions: (1) CC's annotations support the inference that CC intended from the beginning to sail *back* across the ocean to Europe, avoiding any attempt at circumnavigation, and (2) that CC inferred from Marco Polo that trade with the Far East would always be difficult and unprofitable unless a direct route across the Atlantic could be opened up.

[168] **Jalil Sued Badillo.** *Cristóbal Colón y la esclavitud del Indio en las antillas.* San Juan P.R.: Fundación Arqueológica, Antropológica, Histórica de P.R., 1983. 41 pp.

Attempts to identify and document all the recorded instances in which CC participated in the enslavement of the natives of the lands he discovered.

[169] **Juan Gil.** "Las cuentas de Cristóbal Colón," *Anuario Estud Amer* (Seville), 41 (1984): 425-511.

Rpt. *Temas Colombinas* (Seville: Escuela de Estudios Hispano-Americanos, 1986), pp. 83-157 [numbered 1-75]. Examines account books in the Casa de Contratación and those of Santa Clara, treasurer of Hispaniola; the *Libro de Armadas;* royal cedulas and letters; and CC's own notes on his economic affairs and his records of privileges and concessions. Concludes that, no matter how CC may have resented his treatment by the crown, he died a millionaire.

[170] **Juan Gil.** "Tres notas colombinas," *Hist y Bibl Amer* (Seville), 28 (1984): 73-91.

(1) Suggests that Las Casas recorded the date of CC's landfall as 12 Oct, in flagrant contradiction of the other contemporary historians (who all date it 11 Oct), because of a deep personal aversion—rooted in Christian tradition—to the number 11.

(2) Clarifies the curious phrasing in the capitulations of discovery granted to Hojeda and other adventurers who were allowed to explore the South American mainland beginning in 1499: CC's claim to be viceroy of what he discovered was considered to be forfeited because of his failure to follow his orders on the 3rd voyage in 1498, when he discovered South America.

(3) Explores the effects of Diego C's suit against the crown upon events in Central America in 1518.

[171] **Tzvetlan Todorov.** *The Conquest of America: The Conquest of the Other,* tr. Richard Howard. New York: Harper and Row, 1984. 274 pp.

Harper Colophon ed., tr. from the French ed., *La Conquête de l'Amérique: La Question de l'autre* (Paris: Seuil, 1982), 274 pp. A study of the failure of communication between cultures. Sets forth 4 phases in the apprehension of the other: discovery, conquest, love, and knowledge. A semiotic, or deconstructionist, study. CC, central to the "discovery" phase, fails to recognize the native Americans' humanity and reduces them either to "noble savages" or "dirty dogs."

Thesis: writing leads to the powerful and creative habit of improvisation, whereas oral cultures are inhibited by ritual and stultifying organization.

Book limited by T's lack of comprehension of late medieval habits of thought, as when in the introduction he assumes that question-begging is characteristically medieval and proper inductive establishment of premises characteristically modern.

[172] **A. Boscolo.** "Il Genovese Francesco Pinelli," *Pres Ital Andalu I* (Seville: Escuela de Estudios Hispanoaméricanos, 1985), pp. 249-66.

Rpt. in *Saggi su Cristoforo Colombo* (Rome: Bulzoni, 1986), pp. 15-33. Francesco Pinelli was a merchant and banker of Genoese extraction who was already well-established in Spain when CC arrived with references, perhaps, from other Ligurians in Portugal. P. was, with Luis de Santangel, the co-treasurer of the "Santa Hermandad" and a trusted counselor of the king. He was related through marriage to the Centurione, Genoese merchant-bankers with branches throughout Spain. His marriage to María de la Torre enabled him to enter the local aristocracy.

P was in large part responsible for the funding of CC's voyages, both personally and through friends and relatives, and was one of the merchants who actively traded with the Indies.

[173] **Geo Pistarino.** "Il medioevo in Cristoforo Colombo," *Saggi e Documenti* (Genoa), 6 (1985): 451-77.

In spite of what may be said about CC's medieval reading and Genoese mercantile background, he does not have the spirit of a medieval merchant but of a modern man, inspired by technical advances and by the prospect of new horizons.

[174] **Alberto Boscolo.** "Cristoforo Colombo, La Isabela, e il memoriale Torres," *Saggi su Cristoforo Colombo* (Rome: Bulzoni, 1986), pp. 77-89.

A short history of La Isabela from its foundation in late 1493 to its complete abandonment in 1502.

[175] **Alberto Boscolo.** "Fiorentini in Andalusia all'epoca di Cristoforo Colombo," *Saggi su Cristoforo Colombo* (Rome: Bulzoni, 1986), pp. 63-73.

Florentine merchants were very much present in Andalusia at the time of CC, even though few documents of their transactions exist in the Spanish archives because, unlike the Genoese, the Florentines preferred to deal in trust and put little in writing. Some of these merchants, including Giannotto (Juanoto) Berardi, Pietro Rondinelli, Simone Verde, and Francesco dei Bardi were friends of CC. Cf. Consuelo Varela, #179, #181, and #183.

[176] **Juan Gil.** "Los Franciscanos y Colón," *Archivo Hispano-Americano* (Seville), 46, nos. 181-84 (Jan-Dec 1986): 77-110.

Traces the conflict between CC and the Franciscan Cisneros, whose zeal to reform the treatment of native Americans was a major factor in the humiliation of CC in 1500 and in his ouster from Hispaniola. Strongly suggests that CC's decision to be buried at Las Cuevas instead of at a Franciscan house was a final shot against Cisneros by CC. Cf. Juan Manzano Manzano, *Los Pinzones y el descubrimiento de América* (Madrid: Istituto di Cooperazione Iberoamericana, 1988) (#286), Vol.1, Chap.1, where it is suggested that CC probably resented the protection given to M.A. Pinzón by the Franciscans at La Rábida at the end of Voyage 1.

[177] **Jesús Salgado Alba.** "Enigmas historicos en torno al descubrimiento," *Rev Gen Marina* (Madrid), 211 (1986): 317-30.

Proposes to analyze some of the mysteries of the discovery, e.g., the "pre-discovery," the Atlantic crossing, the landfall, and the *Journal*, and looks forward to the "Patronato Doce de Octubre" when he intends to unveil his findings.

Isabel Arenas Frutos

[178] **Stuart B. Schwartz.** *The Iberian Mediterranean and Atlantic Traditions in the Formation of Columbus as a Colonizer.* Twin Cities: Univ. of Minnesota, 1986. 25 pp.

Traces Italian, especially Genoese, colonization in the eastern Mediterranean, and Spanish-Portuguese colonization in the Atlantic Islands (Canaries, Madeira, Azores) as operative precedents for CC's plan to colonize Hispaniola.

[179] **Consuelo Varela.** "El entorno florentino de Cristóbal Colón," *Pres Ital Andalu II* (Bologna: Cappelli, 1986), pp. 125-34.

Traces the extensive and frequently close relationships between CC and these four Florentines in Seville from the late 1480's until CC's death: Juanoto Berardi, Amerigo Vespucci, Simón Verde, and Francisco de Bardi. Cf. Alberto Boscolo. "Fiorentini in Andalusia," #175, and Varela, #181 and #183.

[180] **Consuelo Varela.** "John Day, los Genoveses, y Colón," *Temi Colombiani* (Genoa: ECIG, 1986), pp. 363-371.

Describes evidence in newly discovered documents locating John Day in Seville and Sanlucar de Barrameda in 1499 and 1501. By placing Day in contact with families in Spain among whom CC moved, this new evidence supports previous evidence suggesting a close acquaintance between Day and CC.

[181] **Consuelo Varela.** "Florentines' Friendship and Kinship with Christopher Columbus," *Columbus and his World* (Ft. Lauderdale FL: CCFL, 1987), pp. 33-43.

When CC went to Spain in 1485 he sought help from the Genoese in Seville, but got none. Instead he was befriended by a rising Florentine immigrant, Juanoto Berardi, who helped finance the first voyage and was ruined by his investment in the second voyage. After B's death, CC associated with Riberol, another Genoese who came to Seville after being in Portugal. He was the only Genoese CC associated with familiarly, for the discoverer never forgave the Genoese in Seville who had rebuffed him, and henceforth associated with persons who like himself had had difficulty being accepted. Cf. Alberto Boscolo. "Fiorentini in Andalusia," #175, and Varela, #179 and #183.

[182] **Giuseppe Bellini.** "'... Andavam todos desnudos ...': Alle origini dell'incontro tra l'Europa e l'America," *Columbeis II* (Genoa: DARFICLET, 1987), pp. 181-201.

CC's writings about the first encounters with the new world are infused with the magic which he, as a medieval man, was capable of seeing and embracing: the magic of nature, of the flora and fauna. And though his early concepts about the American world dissolved as people got to know it, his commitment to knowledge is still valid. But CC, a man of his time, is constantly thinking, even while registering the magic, of how he can exploit the trusting savages he has found.

Rina Ferrarelli Provost

[183] **Consuelo Varela.** *Colón y los florentinos.* Madrid: Alianza, 1988. 171 pp.

A study of CC's association with a group of middle-class Florentine merchants throughout his years in Spain. After a chapter describing the background of this group in Seville in the 15th century, V. treats in successive chapters CC's relations with Amerigo Vespucci, Simón Verde, Francisco de Bardi, and Piero Rondinelli. Documentary appendix. Cf. Consuelo Varela, #179 and #181, and A. Boscolo, #175.

B. BIRTH AND EARLY YEARS

[183a] **Giovanni Batista Spotorno.** *Della origine e della patria di Cristoforo Colombo.* Genoa: Frugani, 1819. 247 pp.

The initial modern Genoese account of CC and his family in Liguria.

[184] **M. Staglieno.** *Il borgo di San Stefano ai tempi di Colombo, e la casa di Domenico Colombo.* Genoa: Pellas, 1881. 30 pp.

A detailed reconstruction of the *borgo* where Domenico C. had his residence and shop and where CC lived. Details the streets, shops, and churches whose existence is documented, as well as some of the people who lived in them: shopkeepers, craftsmen, teachers and lawyers.

[185] **M. Staglieno.** *Sulla casa abitata da Domenico Colombo in Genova.* Genoa: Sordo-Muti, 1885. 99 pp.

DC's house is established to be the one on the left of the 1st Carrogio dritto di Ponticello leading from the Sant'Andrea Gate, No. 37. Later, when his son-in-law took over the house, the site was referred to in the title as being *in contrata porte S Andree*.

[186] **Alberto Salvagnini.** "Cristoforo Colombo e i corsari Colombo suoi conteporanei," *Raccolta*, II.3 (1894):127-248.

Discusses whether CC might have been kin to the contemporary pirates called Colombo or Coullon; concludes not. Reviews what was known at the time of the 4th centennial of Columbus the Archpirate Guillaume de Casenove, called Coullon, (died ca. 1482), and of Columbus Junior, perhaps to be identified with a certain "George the Greek."

The whole discussion is confused by the assumption that the seafight that CC may have been involved in off Portugal took place in the mid-1480's, after the death of Guillaume de Casenove. Later evidence, discussed in "Questioni Colombiane," #194, by Giuseppe Pessagno, shows that Casenove attacked a Genoese convoy off Portugal in Aug 1476. "Columbus Junior" has since been identified as Bissiparte.

Reproduces 122 pertinent documents. See also #142.

[187] **Henry Vignaud.** *A Critical Study of the Various Dates Assigned to the Birth of Christopher Columbus; the Real Date, 1451, with a Bibliography of the Question.* London: Stevens & Stiles, 1903. 121 pp.

Reviews the various available evidence. V's decision for 1451 instead of 1446-7 correctly anticipates the conclusive document published by Assereto in 1904 (see #188).

[188] **Ugo Assereto.** "La data della nascità di Colombo accertata da un documento nuovo," *Giorn Storico e Lett Ligur* (Genoa), 5 (1904) 5-16.

Dates CC's birth within a few months of Aug 1451 on the basis of CC's testimony in a Ligurian court in 1479 concerning his failure to secure an order of sugar in Madeira, as agent for the Di Negro firm's office in Lisbon, on behalf of the Centurione firm of Genoa. Prints the document. Morison, *Journals and Other Documents* (New York: Heritage, 1963), #14, prints the "Assereto Document" in English translation, pp. 8-10.

[189] **Celso García de la Riega.** *Colón, español: su origen y patria.* Madrid: Sucesores de Rivadeneyra, 1914. 185 pp.

Furnishes a battery of photocopied documents demonstrating that CC was born in Pontevedra, Galicia, in 1436 or 1437, the son of Domingo de Colón and Susana Fonterosa. CC went to sea in 1451. In 1452 or 1453, the parents emigrated to Genoa with CC's brother Bartholomew. The rest of the family followed them later.

Refuted by R.D. Carbia, #190. The documents furnished by García were studied in the 1920's by a committee of the R. Sociedad de la Historia and rejected as forgeries; see #s 191, 192, 196.

[190] **Rómulo D. Carbia.** *Origen y patria de Cristóbal Colón. Crítica de sus fuentes históricas.* Buenos Aires: Talleres, 1918. 50 pp.

2nd ed. *La patria de Colón, examen Crítica de las fuentes históricas. . . .* Buenos Aires: Talleres, 1923. 70 pp. Reviews systematically (1) the documentary evidence that CC was from Genoa; (2) García de la Riega's case for Castilian (Pontevedran) origin; (3) the supposition that CC's written language is that of a native Castilian; (4) the supposition that the Mayorazgo of 1498 establishes *Colón* as the *Spanish* name of his forebears; (5) the persistent mystery that seems to hang about the discoverer. Conclusion: the case for Spanish origin is based on false documents.

[191] **Angel de Altolaguirre.** *Colón español? Estudio histórico-crítico.* Madrid: Huerfanos, 1923. 89 pp.

Summarizes the findings of Serrano Sanz and Oviedo y Arce, both of whom pronounced as forgeries Celso García de la Riega's documents showing CC to be Galician.

C. E. Nowell

[192] **Angel de Altolaguirre.** "Declaraciones hechas por d. Cristóbal, don Diego, y don Bartolomé Colón acerca de su nacionalidad," *Bol R Acad Hist* (Madrid), 87 (1925): 307-25.

Defends as authentic CC's 1497 will, which twice acknowledges his Genoese birth. Also refutes claims of Spanish birth by CC's designation of himself as a foreigner in letters to the Spanish monarchs. Reprints most of the contemporary accounts of CC's nationality. All call him Genoese.

[193] **Angel de Altolaguirre.** "La real confirmación del mayorazgo fundado por d. Cristóbal Colón," *Bol R Acad Hist* (Madrid), 88 (1926): 330-55. Rpt. Madrid: Tip. de Archivos, etc., 1926. 27 pp.

Pronounces the two copies of the Royal confirmation of CC's Mayorazgo—copies discovered by Alicia Bache Gould—to be conclusive proof of the authenticity of CC's two assertions of his Genoese birth in the Mayorazgo.

[194] **Giuseppe Pessagno.** "Questioni Colombiane," *Atti Soc Ligur Stor Pat* (Genoa), 53 (1926): 539-641.

The Archpirate Guillaume de Casenove, called Coullon (died ca. 1482), attacked a Genoese convoy off Portugal in Aug. 1476; CC may have been shipwrecked, and may have entered Portugal, during this fight.

[195] **Luis Ulloa.** *Christophe Colomb, Catalan: la vrai genèse de la découverte de l'Amérique.* Paris: Maisonneuve, 1927. 404 pp.

A fair example of the wishful thinking that governs nationalistic attempts to take CC's birthplace away from Genoa. For a concise refutation see George E. Nunn's review, *Amer Hist Rev*, 33 (1927-28): 918.

[196] **Angel de Altolaguirre,** et. al. "Informe sobre algunos de los documentos utilizados por don Celso García de la Riega en sus libros 'La Gallega' y 'Colón español,'" *Bol R Acad Hist* (Madrid), 93 (1928), 39-57.

A blistering report by a committee of the Spanish Royal Academy of History. Analyzes 8 passages from three sets of documents used by García de la Riega (cf. #189, and #711). Concludes (1) that the wording of the documents had been systematically revised to make the names correspond to certain Galician documents of the years 1437-1525; (2) that the ink used in revising the documents was modern; (3) that the revisions had all been done by one person; and (4) that therefore these documents on which García rests his two books cannot be admitted to support a serious historical investigation.

[197] **City of Genoa.** *Colombo. Documenti e prove della sua appartenenza a Genova.* Italian-Spanish ed.

Genoa: Istituto d'Arti Grafiche, 1931.

English-German ed., *Christopher Columbus: Documents and Proofs of his Genoese Origin,* 1931. Pp. vii-xxii, "Justification of the Work: its Content and Method." An exhaustive, carefully documented essay containing these sections: "The 'Genoese School'"; "Tradition and criticism"; "The Alleged 'Obscurity' of C's Personality and its Consequences"; "The Documenttion Regarding C. Compared with Fictitious Statements"; "Contents and Method of the Work"; "Further Points: Conclusion."

Contents, pp. 1-288. 1. Attestations of CC's contemporaries; 2. Notarial deeds and other official documents in the Ligurian archives, and information given by CC's companions and acquaintances; 3. CC's personal deeds and the deeds of his relatives and descendants.

A diplomatic presentation with a photo of each document and a translation into German and English. The Italian-Spanish edition is arranged in the same way. Overwhelmingly convincing documentation.

[198] **Roberto Almagià.** "Cristoforo Colombo cittadino genovese," *Boll Soc Geog Ital* (Rome), 69 (1932): 164-69.

Declares that the evidence of CC's Genoese origin in #197 is overwhelmingly cogent and that it should lay the question of his origin to rest permanently.

C. E. Nowell

[199] **G. Giacchero.** "Colombo e i suoi rapporti col Banco di San Giorgio," *Bol Civico Ist Col* (Genoa), no. 1 (1955), pp. 19-27.

Analyzes the strong implications of CC's Genoese origin implicit in his will of 1502, in which he assigns 1/10 of the income from his estate to the people of Genoa, a gift not acknowledged or acted upon by the Bank of St. George, designated as recipient and agent of the gift. Cf. #53.

[200] **Genovesità de Colón.** *Liguria* (Genoa), 39, no. 11 (Nov. 1972).

A compendium of documents, with opinions by current historians.

[201] **Aldo Agosto,** ed. *La sala colombiana dell'archivio di stato de Genova.* Genoa: Cassa di Risparmio, 1974.

Rpt. 1978. Description of 63 leading documents, 1429-1515, basic evidence of CC's birth to a Ligurian family in Genoa, and residence in Genoa and Liguria. Introductory essay on CC's life drawn from P.E. Taviani's account in CC: *la genesi della grande scoperta*, #441.

[202] **Laura Balletto.** "Chio nel tempo di Cristoforo Colombo," *Atti III Conv Internaz Stud Col* (Genoa: CIC, 1979), pp. 175-98.

A picture of Chios as it appeared in the period 1471-76—embracing the years when CC might have been there—based on notarial records published in Philip P. Argenti's *The Occupation of Chios by the Genoese, etc., (1346-1566),* 3 vols. (Cambridge, Eng., 1958).

[203] **Pietro Sanavio.** "Giovenezza di Cristoforo Colombo," *Cristoforo Colombo nella Genova del suo tempo* (Turin: ERI/RAI, 1985), pp. 329-37.

Sums up and interprets known facts and conjectures about CC's ethnic origin, birthplace and date, education, and the occupations of his youth. Extensive bibliography.

[204] **Piero Sanavio, Adriana Martinelli, and Caterina Porcu Sanna,** eds. *Cristoforo Colombo nella Genova del suo tempo.* Torino: ERI/RAI, 1985. 366 pp.

A collection of essays by various hands on Genoa and its culture at the time of CC. The last chaper, Piero Sanavio's "Giovanezza di Cristoforo," is the only one specifically on CC's life. Extensive bibliographies for each chapter.

[205] **Gianfranco Rovani.** *Quinto e Cristoforo Colombo.* Genoa: Edizioni Rovani, 1986.

Proposes Quinto al Mare as CC's birthplace, since CC identified himself as being "de terra rubia" and this "terrarossa" was and still is part of Quinto. Documents, maps, and pictures.

[206] **Aldo Agosto.** "In quale 'Pavia' studiò Colombo?" *Columbeis II* (Genoa: DARFICLET, 1987), pp. 131-36.

Hypothetically identifies the "Pavia" referred to by Ferdinand C as the locale of CC's schooling with a monastery school in the section of Genoa known today as "Paverano." This could explain the C brothers' ability to grasp mathematical and other concepts better than might be expected if they had received only the elementary schooling furnished to sons of weavers and similar artisans.

[207] **Giovanna Petti Balbi.** "La scuola a Genova e Cristoforo Colombo," *Columbeis II* (Genoa: DARFICLET, 1987), pp. 31-36.

About the kind of early education CC might have received in Genoa, a city where learning had practical ends for the most part. Surmises that "Pavia" probably referred to Vicolo Pavia, a side street off "vico dritto di Porticello," one of the many streets that went up to the Porta Soprana in CC's neighborhood.

There were no public schools; usually parents banded together and hired tutors. There is no record that he was apprenticed either to a weaver or a mapmaker, and it is not likely that he was, since he apparently was at sea at fourteen.

[208] **Paolo Emilio Taviani.** *La Genovesità di Colombo.* Genoa: ECIG, 1987. 93 pp.

Rehearses the overwhelming documentation of CC's Ligurian roots and Genoese birth. Chapters deal with the birthplace disputes; with CC's repeated acknowledgment of Genoese birth; with the extensive testimony in ambassadorial letters of the time; with the converging documentation of CC's birthdate in 1451; with evidence about the Ligurian families of CC's father and mother; with the "casa di Colombo" in Genoa; and with the various imputations of non-Ligurian and non-Christian extraction of CC's family, all rejected in favor of evidence that the family is Catholic and Ligurian on both sides as far back as it can be traced.

C. THE PORTUGUESE YEARS

[209] **P. Gaffarel.** "Cristóbal Colón en Portugal," *Rev Geog* (Paris), 28 (1891): 321-32, 409-420; and 29 (1891): 38-45, 118-122, 194-201.

A thorough examination of the documents available in 1891, attempting to establish CC's date of birth, who his father and mother were, what his education was, what his life as a sailor comprised, how he met and married Felipa Moniz, his relationship with Toscanelli and with Martin Behaim, and what other things his experience with Portuguese maritime operations contributed to his Enterprise of the Indies.

Rina Ferrarelli Provost

[210] **Antonio María Freitas and Regina Maney.** *A mulher de Colombo.* Lisbon: Guedes, 1892. 67 pp.

English parallel rendition, A.M. Freitas and Regina Maney, *The Wife of Columbus* (New York: Stettiner, 1893), 50 pp. A study of the antecedents, Portuguese and Italian, of CC's wife Felipa Perestrello Moniz. Chapters on (1) the Moniz family; (2 & 3) the Perestrello family; (4) the marriage of CC to Felipa.

Freitas (who writes under the name Nicolau Florentino) does not mention Maney in the Portuguese version; in the English version Maney claims partnership with Freitas from the beginning of the project, and provides much more detailed genealogical charts.

[211] **Angel de Altolaguirre.** "Llegada de Cristóbal Colón à Portugal," *Bol R Acad Hist* (Madrid), 21 (1892): 481-92.

Concludes there is no reliable evidence that CC was in Portugal before 1476.

[212] **Prospero Peragallo.** *I Pallestrelli di Piacenza in Portogallo e la moglie di Cristoforo Colombo.* Genoa: Papini, 1898. 85 pp.

Disquisizioni Colombiane, 5. A study of the antecedents, both Portuguese and Italian, of CC's wife Felipa Perestrello Moniz. Chaps. 1 & 2 on the Perestrello family; #3 on the Moniz family; #4 on CC's sister-in-law Violante, Briolante, or Michel Muliar or Muliarte, resident of Huelva, Spain (near Palos); #5 on Bartolomeo

Perestrello, Felipa's brother, and his family; #6 on the marriage of CC to Felipa. Note close similarity of structure to #210.

[213] **Charles E. Nowell.** "The Rejection of Columbus by John of Portugal," *University of Michigan Historical Essays* (Ann Arbor: Univ. of Michigan Press, 1937), pp. 25-44.

Rejects Las Casas' account of CC's proposal to John II in 1484 as fabricated from the *Capitulations of Santa Fé* (1492); CC would not have asked John for the privileges accorded to the admirals of Castile. The accounts of the proposal by 3 Portuguese chroniclers agree on Cipangu as the target of the expedition. In rejecting the proposal, the king's regular council on the route to the east was acting as one would expect, in view of CC's vast claims and his dubious proposal.

Mug 71

[214] **Vilhjamur Stefansson.** *Ultima Thule: Further Mysteries of the Arctic.* New York: Macmillan, 1940. 383 pp.

In "Did C Visit Thule?" and "Were Pythias and C Right about Arctic Climate?" S supports the report of Ferdinand Columbus and of Las Casas that CC did so, and that he found no ice in waters 100 leagues north of Iceland.

[215] **Gaston Broche.** "Christophe Colomb a-t-il atteint l'Islande?" *Studi Colombiani* (Genoa: SAGA, 1952), 3: 249-56.

Rejects Ferdinand C's assertion that CC sailed to Iceland.

[216] **Samuel Eliot Morison.** "Christophe Colomb et le Portugal," *Bol Soc Geog Lisboa*, 74 (1956): 269-78.

Reviews what is known of CC's Portuguese experience, 1476-1488, and points out that his first voyage of discovery was a logical consequence of the Portuguese marine achievements and explorations of the 15th century.

[217] **Alejandro Cioranescu.** "Portugal y las cartas de Toscanelli," *Estudios Americanos* (Seville), 14 (1957): 1-17.

The Toscanelli map and letter were forged, not by CC but by Portuguese supporters—perhaps including Martin Behaim—aiding King John II in his attempts to limit Spanish claims in the newly discovered lands. Not only Ferdinand C and Las Casas were fooled by these forgeries, but most modern scholars, too. The forgeries do not support a valid claim against the probity of Las Casas. See, however, #s 528-34, especially #529.

[218] **A. A. Ruddock.** "Columbus and Iceland: New Light on an Old Problem," *Geog Jour* (London), 136 (1970): 177-89.

CC never went to Iceland, and never claimed to have done so in any of his extant writings. But he probably left notes on such a voyage, learned of at the friary of La Rábida, where documentary sources show that Bristol seamen in a ship named *Trinity* visited Huelva in 1480 and no doubt talked with the friars, including perhaps Marchena, who would have passed on to CC information on the Bristol voyages to Iceland and also perhaps the information of the 50-ft. tides, transferred in someone's mind from Bristol (where they occur) to Iceland. This could also account for CC's error, in the account reported by Ferdinand C in the Historie, of placing Iceland in 73 deg. instead of 63 deg. latitude.

[Problem: the pertinent document as reported here has the *Trinity* waiting at Huelva for a shipment of gunpowder to come down the river from Seville. This would have left time, of course, for many sailors' yarns, since Seville is not on the Odiel or the Tinto, but far away on the Guadalquivir. fp]

[219] **A. Teixeira da Mota.** "Colón y los Portugueses," *El viaje de Diogo de Teive. Colón y los Portugueses.* (Valladolid: Casa-Museo de Colón, 1975), pp. 30- 63.

Cuadernos Colombinos no. 5. A survey of what is known.

[220] **Gaetano Ferro.** *Le navigazione lusitane nell'Atlantico e Cristoforo Colombo in Portogallo.* Milan: Mursia, 1984. 252 pp.

An amplified edition of Ferro's 1974 book *I navigatori portoghesi sulla via delle Indie.* The 1984 book places CC's sojourn in Portugal and the discoveries of the South American Atlantic coast within the context of the Portuguese explorations. CC, pp. 98-248.

D. SPANISH YEARS 1485-1492

[221] **José Coll.** *Colón y la Rábida.* Madrid: Lib. Católica de Gregorio, 1891.

2nd, revised ed. (Madrid: Los Huérfanos, 1892), 489 pp. Pioneering work anticipating and laying the plan for Angel Ortega's *La Rábida*, #226. Chapters 19-26 treat of CC, Diego C., and the two important friars Antonio de Marchena and Juan Pérez. Superseded by Ortega's work. See also A. Rumeu de Armas, *La Rábida y el descubrimiento de América: Colón, Marchena, y fray Juan Pérez.* (Madrid: Cultura Hispánica, 1968), #230.

[222] **Pierre F. Mandonnet.** "Christopher Columbus and the Dominican Diego de Deza," *Columbus Memorial Volume* (New York: Benziger, 1893), pp. 57-97.

CC in 1504 told his son Diego that de Deza "was the cause of their Highnesses possessing the Indies" and told him of his own decision to remain in Castile in early 1492.

Mug 60

[223] **E. G. Ravenstein.** "Voyages of Diogo Cão and Bartholomeu Dias," *Geog Jour* (London), 16 (1900): 638-49.

Significant for the date of CC's supposed presence in Lisbon when Dias returned from the Cape of Good Hope.

[224] **Antonio Sánchez Moguel.** "Algunos datos nuevos sobre la intervención de Fr. Hernando de Talavera en las negociaciones de Cristóbal Colón con los Reyes Católicos," *Bol R Soc Hist* (Madrid), 56 (1910): 154-58.

Hernan de Talavera, Archbishop of Granada, on 5 May 1492 donated 140,000 maravedis toward CC's Enterprise of discovery.

[225] **Charles H. McCarthy.** "Columbus and the Santa Hermandad in 1492," *Cath Hist Rev* (Wash. DC), 1 (1915): 38-50.

The S.H., which lent the crown 1,140,000 maravedis to equip CC's fleet, was a league of towns organized in 1295 and revived by Isabel to help institute order in Castile. It was in CC's day a state-directed police force, independent of the regular courts, which could collect money from citizens in Castilian towns.

Mug 55

[226] **Angel Ortega.** *La Rábida: historia documental crítica.* Seville: Editorial de San Antonio, 1925-26. 4 vols.

Volumes 2 & 3, *Epoca Colombina.* A very elaborate reconstruction, with liberal quotes from pertinent documents, of (vol. 2) the career of CC, emphasizing (but not exclusively) the navigator's relationship with the friary; and (vol. 3) the participation of mariners from the Tinto-Odiel area in the discovery and the subsequent explorations. Heavy dependence on *Los Pleitos* for documentation.

This work called severely into question by Antonio Rumeu de Armas in *La Rábida y el descubrimiento de América: Colón, Marchena, y fray Juan Pérez* (Madrid: Cultura Hispánica, 1968), #230.

[227] **Rinaldo Caddeo.** "La preparazione finanzaria della Grande Scoperta e l'opera dei finanziere genovesi in Spagna," *La Grande Genova* (Genoa), 9 (1929): 21-25.

CC was long in friendly contact with the Genoese Francesco Pinelli, treasurer of the Santa Hermandad, from which Luis de Santangel raised funds for the first voyage. It is not true that CC was totally out of his element in Portugal and Spain; he definitely made contact with Genoese. This militates against the nationalistic claims that CC was Portuguese, Catalonian, Gallegan, etc. Also, it appears that CC's extraordinary avidity may have been, for many, proper administrative solicitude for the claims of his creditors.

[228] **Antonio Palomeque Torres.** "Ambiente político y científico que rodeó al futuro Almirante de Indias d. Cristóbal Colón en la España de los Reyes Católicos," *Studi Colombiani* (Genoa: SAGA, 1952), 2: 303-355.

Reviews the Spanish scene under Isabel and Ferdinand, addressing the dynastic war with Portugal; the founding of the Santa Hermandad; the subjection of recalcitrant nobles to the crown; the sovereigns' "Catholicity"; the vigorous assertion of royal prerogatives vis-á-vis the papacy; the Inquisition as an instrument of monarchy; the Spanish advisory and governing councils and royal strategies to get them in harmony; and developments on the economic, cultural, exploratory, imperial, educational, and scientific fronts.

[229] **Juan Manzano Manzano.** *Cristóbal Colón: siete años decisivos de su vida, 1485-1492.* Madrid: Ediciones Cultura Hispánica, 1964. 531 pp.

Traces CC's movements through Spain with special reference to the movements of the court. Argues strongly that CC was informed in Aug. 1487 of the Royal Commission's rejection of his Enterprise, and that CC's 2-year stay with Medinaceli came after he returned from visiting John II in Lisbon in 1488. Extends the work of José de la Torre (#146) on CC's liaison with Beatriz Enríquez de Harana. Demonstrates and documents the fact that CC had his son legitimatized (cf. #622). M. strongly hints his belief in the myth of the unknown pilot, later elaborated in *Colón y su secreto* (1976), #426.

[230] **Antonio Rumeu de Armas.** *La Rábida y el descubrimiento de América: Colón, Marchena, y fray Juan Pérez.* Madrid: Cultura Hispánica, 1968. 180 pp.

A skeptical analysis of prior claims about CC's visits to La Rábida. After introductory essay (ch1) on La Rábida itself, Antonio de

Marchena, and Juan Pérez, develops the following points: (ch2) CC was not at La Rábida in 1485, nor was Marchena; but Marchena was CC's friend and sponsor from very early in CC's stay in Spain.

(Ch3) CC went to the Huelva area in 1491 to leave Diego with the Molyarts while he went on to France. At La Rábida he met Juan Pérez and García Hernández. Fr. Pérez undertook to reopen CC's Enterprise with the sovereigns. (Ch4) Juan Pérez was not the superior of La Rábida, who was an astrologer; but the astrologer was not Fr. Marchena either. CC talked with Pedro Vásquez de la Frontera about the latter's voyaging in the Atlantic. Rumeu identifies Vásquez as the pilot Pedro de Velásquez, native of the Huelva area, who had returned home after long service as a pilot for Portugal.

(Ch5) Pérez negotiated the *Capitulations of Santa Fé* for CC; and he and Pedro Vásquez both influenced M.A. Pinzón to join CC's Enterprise. (Ch6) Related problems: (1) R. denies that the sovereigns rejected a request of M.A. Pinzón to come to Barcelona in 1493; (2) he seriously doubts.that Marchena or Pérez went on the 2nd voyage; (3) he thinks Seville the most likely point of CC's entry into Spain; (4) he rejects all testimony about the Medinas except Medinaceli's letter to Mendoza; (5) he thinks CC's negotiations with the Medinas *preceded* his royal audience. Medinaceli put him in touch with the crown.

[231] **Ernesto Lunardi.** "Le 'Capitulaciones de Santa Fé': Gli interessi dei re cattolici e del genovese nella creazione del titolo di 'Almirante del Mar Oceano'," *Atti II Conv Internaz Stud Col* (Genoa: CIC, 1977), pp. 277-305.

Proposes that CC's successful quest for the admiralcy reflects the continuing desire of the sovereigns to reduce the power of highly placed subjects; in this instance, the Admiral of Castile.

[232] **Antonio Rumeu de Armas.** El *'Portugues' Cristóbal Colón en Castilla.* Madrid: Ediciones Cultura Hispánica, 1982. 92 pp.

Adopts the position that the royal payment of 30 Castilian doblas to an unnamed Portuguese on 18 Oct 1487 was made to CC, whose name was left out deliberately. Proceeds to argue that CC, who found it enhanced his prestige to be known as Portuguese, was actually Genoese, but that his Castilian writing was affected strongly by his long stay in Portugal and on Portuguese ships and possessions. These arguments later developed in the author's "Un documento inédito sobre CC," *Saggi e Documenti* (Genoa), 6 (1985): 435-49, #76.

[233] **Paolo Emilio Taviani.** "Si perfezionò in Castiglia il grande disegno di Colombo," *Pres Ital Andalu I* (Seville: Escuela de Estudios HispanoAméricanos, 1985), pp. 1-19.

Reviews the ideological and practical support CC received from Marchena; his exhibition of a world map to the monarchs, and the resulting purchase of a *Ptolemy* by King Ferdinand; his movements with the court and liaison with Beatriz Enríquez; and the annotations that CC and Bartholomew C made in their copies of D'Ailly's *Imago Mundi*, Pius II's *Historia Rerum*, and Marco Polo's *Milione*.

[234] **Paolo Emilio Taviani.** "Ancora sulle vicende di Colombo in Castiglia," *Pres Ital Andalu I*, (Seville: Escuela de Estudios HispanoAméricanos, 1985), pp. 221-48.

A more detailed review of CC's activities, encounters, and reading in Spain, 1485-92. Cf. #233.

[235] **F. Udina Martorell.** "Las capitulaciones de Colón y problemas que plantea," *Saggi e Documenti* (Genoa), 6 (1985): 427-33.

A paleographic study of the question whether the chancellery copy of CC's *Capitulations* of 1492 was transcribed into the royal register of Aragón, #3569, fols. 135 verso and 136 recto, *after* CC had returned from the first voyage in 1493. Concludes that the *Capitulations* were copied into the register between 17 and 30 April, 1492, in fully authentic fashion, and that the curious phrase "ha descubierto," which has sometimes been read to suggest that CC had already discovered the West Indies when the capitulations were written, is not a posterior addition but is the phrase actually used in the now-lost original of the *Capitulations.* Whatever the significance of the phrase, it was not added later but is an authentic part of the document.

[236] **Alberto Boscolo.** "Gli Esbarroya amici a Córdova di CC," *Pres Ital Andalu II* (Bologna: Cappelli, 1986), pp. 13-19.

Rpt. from *Atti Soc Ligur Stor Pat* (Genoa, 1983), pp. 123-31. Traces CC's relationship with the apothecaries who introduced him to the Haranas and thus put him in contact with Beatriz Enríquez de Harana. Emphasizes B's ability to read and write, and to some degree, perhaps, idealizes her attraction for CC. Points out the Haranas' continued good relations with CC's family, e.g., Beatriz's nephew Diego, who became secretary to Diego C's wife María de Toledo.

[237] **Paolo Emilio Taviani.** "Brevi cenni sulla residenza di Colombo in Andalusia," *Pres Ital Andalu II* (Bologna: Cappelli, 1986), pp. 7-12.

Traces the chief conjectures about CC's whereabouts and activities in Spain, 1485-92. Concludes not only that CC was frequently in contact with, and subsidized by, the Genoese in Andalusia, but that the likeliest period for his activity as a bookseller (and reader-annotator of books) was between Oct 1487 and Mar 1488.

[238] **Foster Provost.** "Columbus's Seven Years in Spain Prior to 1492," *Columbus and his World* (Ft. Lauderdale FL: CCFL, 1987), pp. 57-68.

The large number of unverifiable assertions in major scholarly treatments of CC's life causes widespread confusion about the most crucial period in the discoverer's life. We cannot date or even place in sequential order various events in the mariner's life in these years. The attempt to date and order these events has led biographers to create and describe unverifiable scenarios as if they had really happened. The only hope for establishing sounder points of reference lies in the Spanish archives.

E. THE VOYAGES

1. General Treatments

[239] **Henry Harrisse.** "Chronology of Maritime Voyages Westward, Projected, Attempted, or Accomplished, Between 1431 and 1504," in *The Discovery of North America, A Critical, Documentary, and Historic Investigation* (London: H. Stevens, 1892; rpt. Amsterdam: N. Israel, 1961), pp. 649-700.

Section 1 describes 17 pre-Columbian projects, dated 1431-1486, plus 3 of doubtful date between 1491 and 1498. Sec. 2 describes 81 projects 1492-1504, beginning with CC's first voyage. Executed with Harrisse's usual thoroughness.

Appendix, pp. 363-648, "Cartographia Americana Vetustissima," #488 in this Guide.

[240] **Otto Neussel.** "Los cuatro viajes de Cristóbal Colón para descubrir el nuevo mondo," *El Centenario* (Madrid), 2 (1892): 80-96.

A useful date-by-date sequential account of the four voyages. Unfortunately not keyed to the sources except for the first voyage.

[241] **Leo Wiener.** *Africa and the Discovery of America.* Philadelphia: Innes, 1920-22. 3 vols.

Vol. 1 critically examines CC's 1st voyage on the basis of CC's *Journal* and "Letter to Santangel" and the 2nd on the basis of Bernáldez's and Ferdinand Columbus's quotation from CC's journal. Severely attacks CC's observations on the native Americans and Ramón Pané's remarks on native language (see #242); also attacks (ch.4) CC's observations on the natives' bread as inaccurate. In ch. 3, holds that tobacco cultivation and use are Arabic developments which, if CC's observations are true, had somehow reached Cuba already; also holds that the use of tobacco was not widespread when CC arrived.

[242] **James Williams.** "Christopher Columbus and Aboriginal Indian Words," *Proc Int Cong Amer*, 23 (1928): 816-50.

A detailed attack on Leo Wiener's theory of an African origin for American Indian words (#241). Questions the quality of Wiener's evidence.

[243] **E. Hardison Pizzaroso.** *Colón y Canarias.* Tenerife: 1943.

About Isabel Beatriz Bobadilla.

[244] **John J. Johnson.** "The Introduction of the Horse into the Western Hemisphere," *Hisp Amer Hist Rev*, 23 (1943): 587-610.

Española, to which CC took horses on the 2nd voyage in 1493, became the first home of the horse in the Western Hemisphere. From there the Spanish took horses to Puerto Rico, Jamaica, and Cuba; and from these islands was drawn the horse population of South and Central America and the southwestern United States. Details on CC's involvement, pp. 588-96.

[245] **Leonardo Olschki.** "The Columbian Nomenclature of the Lesser Antilles," *Geog Rev* (New York), 33 (1943): 397-414.

Examines the problems involved in identifying the islands CC discovered and named in the Lesser Antilles.

[246] **Francisco Domínguez Compañy.** *La Isabela: primera ciudad fundada por Colón en América.* Havana: Sociedad Colombista Panamericana, 1947. 72 pp.

A history. Part 1: Foundation of the first Spanish-American city. Chapters: 1. Selection of the site. 2. Construction of the city. 3. Constitution of the local government. 4. Social condition of the first citizens. Part 2: First Steps. Chapters: 1. Work, illness, and misery.

2. Economic disaster. 3. First Spanish-American civil war (Roldán rebellion). 4. Social and economic aspect of the Indian question (i.e., slavery of the Indians). Part 3: Agony and Death of La Isabela.

[247] **Alejandro Cioranescu.** *Colón y Canarias.* Laguna de Tenerife: Instituto de Estudios Canarias, 1959. 227 pp.

Ch. 1. There is no reliable evidence of any presence by CC in the Canaries prior to Voyage 1. Chs. 2-3 summarize CC's stops in the Canaries on the remaining voyages. Ch. 4. CC's visits to Gran Canaria. Ch. 5. Reviews what is known of Beatriz de Bobadilla y Peraza, and minimizes the probability of any significant romantic relationship with CC. Ch. 6. CC perhaps stopped on the 4th voyage to see the then governor, Antonio de Torres, who is the person in the Canaries most likely to have had a personal relationship with CC. CC had entrusted the command of his returning fleet to Torres early in 1494, and Antonio's sister Juana Velásquez de Torres was CC's close friend, to whom he wrote a well-known letter detailing the injustices he received at the hands of the royal agent Francisco de Bobadilla. Ch. 7. Importance of the Canaries as the launching point for expeditions to the Indies. Ch. 8. Review of possible or probable relationships between CC and other Canary residents.

[248] **Samuel Eliot Morison and Mauricio Obregón.** *The Caribbean as Columbus Saw It.* Boston: Little, Brown, 1964. 252 pp.

An account and photographic record of flights taken by the authors in 1963 to view all the locales visited by CC in the Caribbean in his four voyages. Voyage 1, 1492-93, Bahamas, NE Cuba, north coast of Hispaniola, Chaps. 1-6. Voyage 2, 1493-96, Leeward Islands, Virgin Islands, Puerto Rico, Hispaniola, south coast of Cuba, Jamaica, Chaps. 4-7 and 11. Voyage 3, 1498, Trinidad, Gulf of Paria, Margarita, southern Hispaniola, Chaps. 5 & 8. Voyage 4, 1502-4, Windward Islands, Hispaniola, Mainland Honduras to Gulf of Darien, Jamaica, Chaps. 5,6,9,10,11.

[249] **R. Barreiro Meiro.** "Juan de la Cosa y su doble personalidad," *Rev Gen Marina* (Madrid), 179 (1970): 165-91.

Documented argument which concludes that Juan de la Cosa, the master of the *Santa María* on the 1st voyage, the mapmaker and mariner of the *Niña* on the 2nd, the companion of Ojeda and Bastidas on the 3rd, and the maker of the world map dated 1500 are all one person. The theory that more than one person named Juan de la Cosa was involved in these activities stems from Alicia Bache Gould's misreading of a royal cedula of 25 August 1496. This led her, and after her Morison, into believing there was more than one Juan de la Cosa.

[250] **Francisco Morales Padrón.** *Sevilla, Canarias, y América.* Las Palmas: Cabildo Insular, 1970. 356 pp.

Explores the symbiotic relationship among the Canaries, the New World, and the chief port of embarcation in Spain, beginning with 1485, the year when CC and the Canarian king Guanarteme both went to Castile to seek aid from the monarchs. In the chapters on CC, pp. 39-53, emphasizes that the plantains and sugar cane that were immediately and successfully planted in Hispaniola both came from the Canaries.

[251] **Samuel Eliot Morison.** *The European Discovery of America: The Southern Voyages.* New York: Oxford Univ. Press, 1974. 758 pp.

Pp. 3-183 and 236-271 provide a succinct, updated paraphrase of *Admiral of the Ocean Sea*, with updated notes.

[252] **Paolo Emilio Taviani.** *I viaggi di Colombo: la grande scoperta.* Novara: Istituto Geografico de Agostini, 1984. 2 vols.

Sequel to CC: *La genesi della grande scoperta,* #441. Organized in the same way: Vol 1, 40 chapters comprising a narrative account of CC's career from the beginning of the first voyage to CC's death, interspersed with essays on pertinent topics, e.g., Ch.17, "The Cannibals," Ch. 33, "Pearls." Vol. 2, 40 chapters keyed to those in Vol. 1, each containing one or more long scholarly notes or "schede" and an extensive bibliography of the subjects treated in the schede.

[253] **Anna Unali.** "L'oro nei primi due viaggi di Cristoforo Colombo nelle Indie," *Temi Colombiani* (Genoa: ECIG, 1986), pp. 317-343.

Describes the gradual shift in the attitude toward the West Indian natives taken by the discoverers, who under the pressures of the desire for power and wealth no longer viewed the natives as admirable but as appropriate subjects for enslavement.

[254] **Aldo Albonico.** "Il malgoverno dei Colombi all'Hispaniola," *Columbeis II* (Genoa: DARFICLET, 1987), pp. 203-223.

A review and analysis of the documentary sources on the subject, which are scarce, and of the early secondary accounts written at the time or shortly after 1493-1500, but not including FC's *Historie*, Las Casas' *Historia*, or Gómara's *Historia general*.

[255] **Kathleen Deagan.** *Artifacts of the Spanish Colonies of Florida and the Caribbean, 1500-1800. Vol. 1: Ceramics, Glassware, and Beads.* Washington: Smithsonian Institution Press, 1987. 222 pp.

The terrestrial sites treated include La Isabela, founded 1493 by CC; Concepción de la Vega, i.e., Vega Real, in central Dominican

Rep., founded ca. 1495; and Sevilla Nueva, Jamaica, at the site inhabited at St. Ann's Bay, beginning 1502, by CC and his stranded seamen on the 4th voyage. The city of Santo Domingo, founded 1498, is a "multi-component site." All four sites are cited *passim* throughout the book. See Index.

2. The First Voyage

a. General

[256] **Adolfo De Castro.** "Los Pinzones," *El Centenario* (Madrid), 1 (1892): 271-84, 320-32.

A defense of Martín Alonso Pinzón against historians who have echoed CC's hard words about him, and a highly complimentary review of Vicente Yañez Pinzón's career.

[257] **Cesáreo Fernández Duro.** "Tripulación de la nao 'Santa María' y de los carabelas 'Pinta' y 'Niña,' con noticias breves de personas y naves en los viajes de Cristóbal Colón," *El Centenario* (Madrid): 1 (1892): 483-89.

A first effort to assemble the information that Alicia Bache Gould brought together, by exemplary research, in # 260.

[258] **Alicia Bache Gould.** "Nuevos datos sobre Colón y otros descubridores: Datos nuevos sobre el primer viaje de Colón," *Bol R Acad Hist* (Madrid), 76 (1920): 201-214.

The pardons of 4 members of CC's crew on Voyage 1, viz., of Bartolomé de Torres, Alonso Clavijo, Juan de Moguer, and Pero Izquierdo. The last three helped Torres escape from jail prior to embarcation. All four were pardoned by the monarchs in May 1493. Implies that these were the only criminals in the 1492 fleet.

This first installment in the long series of documents located and published by Alicia Gould anticipates by 4 years the resumption of the series, renamed "Nueva lista documentada de los tripulantes de Colón en 1492," in *Bol R Acad Hist*, 85 (1924): 34-49. For the remaining information, see #260.

[259] **Luis Morales Pedroso.** *Lugar donde Colón desembarcó por primera vez en Cuba.* Havana: Soc. Geog. de Cuba, 1923. 97 pp.

Analyzes the *Diario* and concludes that CC's landfall on Cuba was at Gibara.

[260] **Alicia Bache Gould.** "Nueva lista documentada de los tripulantes de Colón en 1492," *Bol R Acad Hist* (Madrid), 85 (1924): 34-49.

Resumes the series begun as "Nuevos datos sobre Colón y otros descubridores," *Bol R Acad Hist,* 76 (1920): 201-214. The whole series was collected posthumously, and published as *Nueva lista documentada de los tripulantes de Colón en 1492,* ed. José de la Peña y Camara. Madrid: Real Academia de la Historia, 1984. 551 pp.

This monumental publication, with its wealth of information not only about CC and his 1492 crew but about a host of related matters, probably constitutes the greatest piece of CC research in the 20th century. Analysis: Pp. 1-30, a life and eulogy of Alicia B. Gould by Ramón Carande, dedicated to Ursula Lamb. The remainder of the book draws together all the articles bearing this title in the *Bol R Acad Hist,* 1924-73, and adds another: the article "Nuevos datos," #250, above. The 1973 article (#272) was synthesized posthumously by José de la Peña from Miss Gould's notes.

The series: 1. "Nuevos datos," (1920), above; and the following items entitled "Nueva lista documentada, etc.": 2. 85 (1924):145-59; 3. 85 (1924):353-79; 4. 86(1925):491-532; 5. 87(1925): 22-60; 6. 88(1926):721-84; 7. 90(1927):532-60; 8. 92(1928):776-95; 9. 110(1937-42):91-161; 10. 111(1942):229-90; 11. 115(1944):145-88; 12. (ed Peña) 170(1973):237-317.

[261] **E. Ward Loughran.** "Did a Priest Accompany Columbus in 1492?" *Cath Hist Rev* (Wash DC), 16 (1930-31), 164-74.

The evidence shows no priest, no masses, no confession aboard the ships.

Mug 52

[262] **Glenn Stewart.** "San Salvador Island to Cuba: A Cruise in the Track of Columbus," *Geog Rev* (NY), 21 (1931): 124-30.

Stewart sails a small boat on the course Watlings – Rum Cay – Long Island – Bird Rock and Crooked Island – Nipe Bay, Cuba. Emphasizes the accuracy of CC's descriptions.

Mug 93

[263] **William B. Goodwin.** *The Lure of Gold, Being the Story of the Five Lost Ships of Christopher Columbus.* Boston: Meador, 1940. 213 pp.

Rejects Limonade-sur-Mer as the site of Navidad, and proposes Caracol Bay instead.

Mug 34

[264] **Samuel Eliot Morison.** "The Route of Columbus along the North Coast of Haiti, and the Site of Navidad," *Trans Am Phil Soc* (Philadelphia), New series, 31 (1940), 239-85.

Rpt. Philadelphia: Am Phil Soc, 1940. 47 pp. Poses three main problems: the site of the wreck of the *Santa María*, the site of Navidad, and the site of Guacanagarí's village. M. does not distinguish betwen the royal village some miles away and the village at the site of Navidad, also subject to Guacanagarí. Traces CC's course for each day in voyage from 5 Dec 1492 to 16 Jan 1493. Concludes that Navidad must be within 1/2 mile of the church at Limonade Bord-de-Mer.

[265] **Charles F. Brooks.** *"Two Winter Storms Encountered by Columbus in 1493 near the Azores,"* *Bulletin of the American Meteorological Society* (Milton MA), 22 (1941): 303-309.

Discusses and diagrams the storms of 12-15 Feb and 27 Feb – 4 Mar, which closely resemble the storms of 1936 and 1937 in the same area. Concludes that the ones CC encountered were not simple circular storms but storms marked by "well developed fronts."

Mug 15

[266] **Leonardo Olschki.** "What Columbus Saw on Landing in the West Indies," *Proceedings of the American Philosophical Society* (Philadelphia), 84 (1941), 633-59.

In the *Journal* CC's vague descriptions of natural features, like those of Dr. Chanca and Vespucci, contrast with his realistic accounts of human native Americans, and reflect the primary interest of the age in human beings, a continuation of the medieval split between empirical record and theoretical speculation. CC's *Journal* (not composed, as was the "Letter to Santangel," to impress the monarchs) is full of self-deceptions passed down to us in Las Casas' honest, unmutilated abstract. This abstract reflects both CC's visionary expectations and his occasional recognition of the reality before his eyes.

[267] **William Herbert Hobbs.** "The Track of the Columbus Caravels in 1492," *Hisp Amer Hist Rev*, 30 (1950), 63-73.

Also in *Michigan Alumnus Quarterly Review*, 56 (1950): 118-25. Applies corrections to the compass bearings of CC's *Journal*, derived from US Navy figures on compass variation for 1945 (close to those of 1492). Concludes on the basis of these corrections and on inferences drawn from bird life cited in the *Journal* and CC's remarks about the winds, that CC's track drops far south of that plotted by Morison through the North Tropical Zone of Calms, remaining on the southern fringe of the Sargasso sea, and swinging north again along the lesser Antilles and into the Bahamas. Interesting abstract of statements in the *Journal* relating to the sea log and others that help to map the course.

[268] **Ettore Remotti.** "Concetti antropologi de Cristoforo Colombo," *Studi Colombiani* (Genoa: SAGA, 1952), 3: 243-47.

Praises CC's skillful record of his first impressions of the Bahaman natives, a record that constitutes a large part of our knowledge of these people fated to quick extinction in the early 16th century. Emphasizes the poignant contrast between these Amazonnids and another branch of Amazonnids, the Caribs, whose fierce aggressiveness injects an aura of terror into CC's account. Ends with ironical references to the Bahamans' fate at the hands of "Christian and civilized" Europeans.

[269] **Francisco Morales Padrón.** "Las Relaciones entre Colón y Martín Alonso Pinzón," *Rev Indias* (Madrid), 31 (1961): 95-106.

A brief, straightforward account of Pinzón's relationship with CC, 1492-93, based chiefly on CC's *Journal* and *Los Pleitos*. Credits Pinzón with keeping a despairing CC from giving up the 1st voyage before arrival in the Bahamas.

[270] **Demetrio Ramos Pérez.** *Posible explicación de la escala de Colón en las Canarias.* Tenerife: Aula de Cultura, 1962. 37 pp.

3 reasons: (1) technical superiority of the Canaries as a staging area for exploration in the Atlantic; (2) the personal experience of CC leading him to view these islands, especially Gomera, as a key point in his Enterprise of the Indies; (3) the political fact that Castilian domination of the Canaries signified Castilian domination of the Atlantic and the lands beyond it.

Juan Gil and Bermejo García.

[271] **Juan Manzano Manzano.** *Los mótines en el primer viaje colombino.* Valladolid: Casa-Museo de Colón, 1971. 45 pp.

Cuadernos Colombinos no. 1. Analyzing various contemporary testimony, concludes that CC faced two mutinies on the first voyage: one on 6 Oct 1492 and the other one or two days before the landfall on Guanahani. The first was smothered with the aid of the Pinzóns. At this point CC felt he could not avoid revealing his "secret" about the Indies to the Pinzóns; this restored their confidence, and CC was therefore able to put down the mutiny of 10 Oct. M infers, moreover, that the unknown pilot who discovered the secret was CC himself.

Martin Torodash

[272] **José M. de la Peña y Camara.** "Nueva lista documentada de los tripulantes de Colón en 1492. Advertencia preliminar," *Bol R Acad Hist* (Madrid): 170 (1973): 237-317.

Announces the forthcoming publication (achieved in 1984, #260) of the whole series of documents with this title published by Alicia Bache Gould, and adds a final, posthumous installment based on Miss Gould's research.

[273] **Stephen J. Greenblatt.** "Learning to Curse: Aspects of Linguistic Colonialism in the Sixteenth Century," *First Images of America* (Berkeley: Univ. Cal Press, 1976), 2:561-80.

Significant vis-à-vis CC's belief as registered in his *Journal* that he could understand what the Indians told him. Refutes two widespread beliefs among Europeans at the time of the early discoveries, i.e., "that Indian language was deficient or non-existent and that there was no serious language barrier" between Indians and Europeans (p. 574). Points out that only much later, at the time of Vico, did Europeans begin to recognize that "each language reflects and substantiates the specific character of the culture out of which it springs" (p. 576).

[274] **Rudolf Hirsch.** "Printed Reports on the Early Discoveries and Their Reception," *First Images of America* (Berkeley: Univ. Cal Press, 1976), 2: 537-39.

Examines 119 items printed between 1493 and 1526, and re-editions of 5 of these items, 1527-32. Concludes that a definite and broadly based interest existed, though publications were largely directed to the better-educated. One-sixth of the reports concern CC; one half, Vespucci; the other third concern all the other explorers.

[275] **James Snyder.** "Jan Mostaert's West Indian Landscape," *First Images of America* (Berkeley: Univ. Cal Press, 1976), 1: 495-502.

Mostaert's early 16th-c painting of an Edenic locale showing a community of naked natives resisting organized Spanish invaders "could represent . . . the campaigns of Columbus in 1492-93 [i.e., 1494-96?], Cortés in 1519, De Soto in 1539, Pizarro in 1531, or Coronado in 1540. But this . . . depiction . . . best reflects the accounts of Coronado's expedition" (pp. 496-97).

[276] **Charles H. Talbot.** "America and the European Drug Trade," *First Images of America* (Berkeley: Univ. Cal Press, 1976), 2:833-44.

Although the direct influence of drugs found in the New World on European medicine was negligible except for the use of chinchona derivatives, the false claims made for heavily imported remedies led gradually to a scientific approach to therapeutics. T. uses the syphilis epidemic, attributed to effects of CC's 1st voyage, as a classic example: the enormous demand for the supposed cure, guaiacum, enriched the importers of this wood but ultimately led to experimentation with various cures and organized study of the results.

[277] **Robert M. Rose.** "The Rock of Sintra: Columbus's Landfall," *Mariner's Mirror* (Greenwich, Eng.), 63 (1977): 227-32.

Identifies the spectacular white rock in Cornelius Athenizoon's painting "Infanta Dona Beatriz arriving at Villefranche in 1552" as the Rock of Sintra, which CC recognized on the morning of 4 Mar 1493 as a storm blew the *Niña* toward the Portuguese coast. Identifies the towers of the "Castello dos Mouros" near the upthrust of the rock. In the foreground appears the chapel of the Monasterio de Gerónimo, which is in a proper position for a sighting from a point on the south bank of the Tagus just south of Lisbon. The ships in the foreground are obviously in the Tagus southwest of Lisbon.

The painting provides a striking image of the rock whose appearance saved CC in the storm by showing him where he was.

[278] **Fernando Morais do Rosario.** "A escala de Colombo em Lisboa na viagem de descobrimento do novo mondo," *Atti III Conv Internaz Stud Col* (Genoa: CIC, 1979), pp. 457-466.

Suggests that CC's landing in the port of Lisbon on his return from the 1st voyage was deliberate, and that this proves CC's close relationship with King John II, whom he would certainly have avoided had he not been confident of a warm reception. Cites the *Journal* entry of 11 Mar, which includes suggestions of warmth and cordiality between CC and both King John and Queen Leonor, as well as CC's wife's relative Don Martin de Noronha, who conducted CC on the trip up the Tagus to visit the king.

[279] **Emilio de la Cruz Hermosilla.** "Los marinos de Colón," *Rev Gen Marina* (Madrid), 211 (1986): 259-64.

Celebrates the achievement of Alicia Bache Gould (#260), who through absolute dedication identified 87 of CC's crewmen on the 1st voyage and the ship each sailed on. Reprints Gould's lists.

Isabel Arenas Frutos

[280] **Angel Díaz del Río Martínez.** "Derrotas de las naves de Cristóbal Colón en las islas Canarias en el viaje del descubrimiento," *Rev Gen Marina* (Madrid), 211 (1986): 303-315.

A study, based on a wide variety of documents, setting forth the possible routes followed by the ships in CC's fleet from Palos to

San Sebastian de Gomera. This study was the basis of the conference "Jornadas sobre las derrotas colombinas entre las islas Canarias."

Isabel Arenas Frutos

[281] **Gian Marco Ugolini.** "Il primo viaggio di Colombo alla scoperta dell'America: una valutazione del costo dell'impresa," *Temi Colombiani* (Genoa: ECIG, 1986), pp. 309-315.

Estimates cost of first voyage, in modern equivalents, at between 1 billion and 3 billion lire, i.e., between 1 million and 3 million dollars at 1000 lire per dollar.

[282] **Kathleen Deagan.** "Initial Encounters: Arawak Responses to European Contact at the En Bas Saline Site, Haiti," *Columbus and his World* (Ft. Lauderdale FL: CCFL, 1987), pp. 229-35.

Summarizes the results of archaelogical research at the En Bas Saline site in Haiti, believed to have been the village of Guacanacaric, the Arawak cassique who asisted CC and his men after the *Santa María* wrecked 24 Dec 1492. CC established La Navidad in G's town, where 39 Spanish men lived in daily contact for nearly 9 months. The research has verified the late 15th-c date of the site, provided information about Arawak society on the eve of European contact, and yielded preliminary data about changes in the Indian community between 1492 and ca. 1515.

KD

[283] **Gaetano Ferro.** "Columbus and his Sailings, According to the 'Diary' of the First Voyage: Observations of a Geographer," *Columbus and his World* (Ft. Lauderdale FL: CCFL, 1987), pp. 99-113.

(1) In the first crossing CC depended primarily on the naked eye, making approximate estimate of distances which he reported on one or more nautical charts. (2) Since all the distances were overestimated, the reduced distances that CC reported to the crew ironically turned out to be close to the real distances. (3) The use of instruments was limited, and CC did not use the "taoleta de marteloio" to determine latitudes. His errors in latitude were intentional, determined by political reasons, to conceal from the Portuguese the true position of lands discovered. (4) It is fairly certain that CC landed at Watlings Island. (5) CC's use of nautical and geographic terms depends largely on the Portuguese and the *lingua franca.*

[284] **Richard Rose.** "Lucayan Lifeways at the Time of Columbus," *Columbus and his World* (Ft. Lauderdale FL: CCFL, 1987), 321-39.

CC's *Journal* provides the only known eyewitness account of Lucayan lifeways. Archaeological work at Pigeon Creek on

Watlings and other prehistoric settlements in the central Bahamas has provided important information on Lucayan origins, subsistence, technology, and trade. A reconstruction of these lifeways helps verify some of CC's observations, and improves our understanding of the Lucayans.

RR

[285] **Irving Rouse.** "Origin and Development of the Indians Discovered by Columbus," *Columbus and his World* (Ft. Lauderdale FL: CCFL, 1987), pp. 293-312.

Ethnohistorians have classified the natives of the West Indies into 3 major groups, Guanahatabey, Taino, and Island-Carib. CC met only Tainos during his first voyage. They spoke a single language, also known as Taino, and shared the same culture, which reached its highest development in Hispaniola and Puerto Rico.

Linguists have assigned the Taino language to the Arawakan family and have traced that family back to the middle of the Amazon Basin by reconstructing its ancestral languages. They find that speakers of its proto-Northern language moved into the West Indies from the Guiana coast about the time of Christ. The Proto-Northerners developed the Taino language after reaching the Greater Antilles, and carried it into the Bahamas.

Archaeologists have confirmed the linguists' conclusions. They have assigned the pottery of the Taino Indians to an Ostionoid series of styles and have traced that series back to a Saladoid series, which originated in the Orinoco Valley. They find that the Saladoid potters entered the West Indies about the time of Christ, introducing not only pottery but also agriculture and zemiism, the religion of the Tainos. The Saladoids and their Ostionoid descendants gradually pushed the previous inhabitants of the islands back into western Cuba, where they became the Guanahatebeys.

The Ostionoids developed Taino culture after reaching the Greater Antilles, and carried it into the Bahamas.

IR

[286] **Juan Manzano Manzano and Ana María Manzano Fernández-Heredia.** *Los Pinzones y el descubrimiento de América.* Madrid: Ediciones Cultural Hispánica, 1988. 3 vols.

An exhaustive, sumptuously documented study of the activities of Martín Alonso Pinzón and Vicente Yañez Pinzón in the discovery and opening up of the New World. Vol. 1, Chap. 1, pp. 5-200, which treats the Pinzóns' association with CC at the time of the 1st voyage, is based on the doctoral thesis of A.M. Manzano Fernándo-

Heredia at the University of Seville. The conclusion about the death of M.A. Pinzón—that his death in Palos may well have been due to syphilis contracted in Hispaniola after he had deserted the fleet on 21 Nov 1492—is an index of the objectivity and impartiality of the scholarship.

b. Landfall at Guanahani

[287] **Mark Catesby.** *The Natural History of Carolina. London*, 1731-47. 2 vols.

3rd ed. (1771), vol. 2, p. xxxviii. Identifies Guanahani as Cat Island. John Parker (#315) designates this as the earliest known attempt to identify the landfall island.

[288] **Samuel Kettell,** ed. *Personal Narrative of the First Voyage of Columbus.* Boston: Wait, 1827.

P. 34, identifies Guanahani as Grand Turk, supporting Navarrete's 1825 opinion (#2).

John Parker

[289] **Alexander S. Mackenzie.** "Appendix" to Washington Irving, *A History of the Life and Voyages of Christopher Columbus* (New York: Carvill, 1828), 3: 207-226.

Refutes Navarrete's 1825 contention (# 2) that Guanahani is Grand Turk Island. Favors Cat Island (like Irving, vol. 2, p. 156).

[290] **George Gibbs.** "Observations to Show that the Grand Turk Island, and not San Salvador, Was the First Spot on Which Columbus Landed in the New World," *Proceedings of the New York Historical Society,* 1846, pp. 137-48.

Attacks the designation of Cat Island as Guanahani.

John Parker

[291] **R. H. Major.** *Select Letters of Christopher Columbus, With Other Original Documents, Relating to his Four Voyages to the New World.* Hakluyt Society, 1st series, 2 (London: Hakluyt Soc., 1847), pp. liii-liv.

2nd ed., Hakluyt Soc. 1st series, 43 (London, 1870), pp. liii-liv. Under influence of George Gibbs, favors Grand Turk as Guanahani.

John Parker

[292] **A. B. Becher.** *The Landfall of Columbus on his First Voyage to America
. . . also a Chart Showing his Track from the Landfall to Cuba*, etc.
London: Potter, 1856. 376 pp.

Dismisses arguments for Turks and Cat Islands, and votes for
Watlings. Detailed analysis of track from Watlings to Cuba.

[293] **G. V. Fox.** *An Attempt to Solve the Problem of the First Landing Place of
Colulmbus in the New World.* Washington DC: Govt. Printing Office,
1882. 68 pp.

Guanahani is Samaná Cay.

[294] **J. B. Murdock.** "The Cruise of Columbus in the Bahamas," *United
States Naval Institute Proceedings* (Annapolis MD), 11 (1884): 449-86.

Guanahani is Watlings.

[295] **Philipp J. J. Valentini.** "The Landfall of Columbus at San Salvador,"
Proceedings of the American Antiquarian Society (Worcester MA),
New Series, 8 (1892): 152-68.

Argues for Cat Island because this would permit the light seen on
the evening of 11 Oct 1492 to be placed on Watlings.

Mug 97

[296] **E. A. D'Albertis.** "Sulla traccia del primo viaggio di Cristoforo
Colombo verso l'America," *Boll Soc Geog Ital* (Rome), Ser. 3, 6
(1893): 741-51.

A letter from the captain of the Ship Corsaro to Giacomo Doria,
describing the Sargasso Sea and Watlings Island as encountered on
the 1893 voyage to New York.

[297] **Jacques W. Redway.** "The First Landfall of Columbus," *National
Geographic* (Wash. DC), 6 (Dec 1894), pp. 179-92.

Supports G. V. Fox's identification of Guanahani with Samaná Cay
(#293). Introduces evidence from maps, especially a world map in
the British Museum painted "par ordre de Henri II."

Mug 84

[298] **Clements R. Markham.** "Sul Punto de Approdo di Cristoforo
Colombo," *Notizie e studi in conessione colla Raccolta pubblicata dalla
R. Commissione Colombiana* (Rome: Soc Geog Ital, 1894), pp. 12-35.

Rpt. *Boll Soc Geog Ital* (Rome), 12 (1899): 101-124. Lists the evidence
from CC's *Journal* as to the first five places encountered, i.e., San

Salvador (Guanahani), Concepción, Fernandina, Isabela (Samoet) and the Islas de Arena. Reports the hypotheses of Muñoz (1793), Watlings; Navarrete (1825), Grand Turk; Washington Irving (1828) and Humboldt (1837), Cat Island; Varnhagen (1864), Mayaguana; Fox (1882), Samana Cay; Becher (1856), Watlings; Murdock (1884), Watlings. Concludes that Muñoz was right (Watlings) ; that Major (1871) established the location of the first anchorage on the southeast [sic] coast (p. 31); and that Murdock (1884) established the route to Cuba from San Salvador through the Bahamas.

[299] **Theodore H. N. de Booy.** "On the Possibility of Determining the First Landfall of Columbus by Archaelogical Research," *Hisp Amer Hist Rev*, 2 (1919): 55-61.

Proposes (1) search for the peninsula in the Bahamas that CC in the *Journal* entry of 14 Oct called "a piece of land like an island, although it is not one, with six houses on it," and (2) an archaeological analysis of the remains of these houses.

Mug 56

[300] **H.C.F. Cox.** "The Landfall of Columbus," *Geog Jour* (London), 16 (1926) : 332-39.

Prints the act of the Bahamas parliament ("16 and 17 Geo V. Chapter 27, assented to 6th May 1926") which legislates the name of Watlings Island to be "San Salvador," since "It has been definitely proved that the landfall of Columbus in the New World was on that island . . . ."

[301] **R. T. Gould.** "The Landfall of Columbus: An Old Problem Re-stated," *Geog Jour* (London) 17 (1927): 403-429.

Reviewing in turn the long series of conjectures about the identity of Guanahani, G acknowledges that the evidence of the *Journal* as abstracted by Las Casas is too contradictory for anyone to achieve conclusive proof; but he is strongly inclined to feel that the weight of all the evidence and arguments makes Watlings the best choice.

[302] **R.T. Gould.** "Decision for Concepcion Cay," *Enigmas* (London, 1945), pp. 66-94.

Rpt. (New Hyde Park, N.Y.: University Books, 1965), pp. 66-94. Rejects Morison's decision for Watlings. Cf. #301.

[303] **E. Roukema.** "Columbus Landed on Watlings Island," *American Neptune* (Salem MA), 19 (1949): 79-113.

Examines the claims for Caicos, Cat, Grand Turk, Mayaguana, Samana Cay, and Concepcion, and rejects them in favor of Watlings.

[304] **Pieter Verhoog.** "Columbus landed on Caicos," *United States Naval Institute Proceedings*, 80 (1954): 1101-1111.

Develops the argument of his earlier presentation, *Guanahani Again. The Landfall of Columbus in 1492* (Amsterdam: C. de Boer, 1947). Only the Caicos Islands can fit satisfactorily into the pertinent remarks identifying and locating Guanahani in the Journal. Attacks as indefensible the accepted status of Watlings as Guanahani.

[305] **E. A. Link and M. C. Link.** *A New Theory on Columbus's Voyage Through the Bahamas.* Wash. DC: Smithsonian Inst., 1958. 45 pp.

Smithsonian Misc. Collections, v. 135, no. 4. Rejects Morison's argument for Watlings, and accepts Verhoog's for Caicos [#304; see V's later restatement of his theory, #320]. Based on the Links' 1955 testing of the two arguments by plane and boat. The Links' candidates for the next 3 islands, Samaná Cay, Long Island, and Crooked Island.

John Parker

[306] **Robert H. Fuson.** "Caicos: Site of Columbus's Landfall," *The Professional Geographer* (Wash. DC), 13, no. 2 (Mar 1961): 6-9.

Argues that the details of CC's *Journal* for the voyage from Guanahani to Cuba fit perfectly if Guanahani is Caicos, but do not fit at all if it is Watlings.

[307] **Edwin Doran, Jr.** "This Columbus-Caicos Confusion," *The Professional Geographer* (Wash.DC), 13, no. 4 (July 1961): pp. 32-34.

Addresses 5 errors that he says R.H. Fuson makes in reading the evidence in "Caicos, Site of C's Landfall," #306. Argues that Morison's analysis in *AOS* "has fewer unexplained facts than any other" (p. 32).

[308] **Robert H. Fuson.** "Caicos, Confusion, Conclusion," *The Professional Geographer* (Wash. DC), 13, no. 5 (1961): 35-37.

Rebuts in order Doran's claims in #307 of 5 errors in reading the evidence in Fuson's previous note "Caicos: Site of C's Landfall," #306. Concludes that Guanahani is still Caicos, as he said before.

[309] **Ruth G. Durlacher Wolper.** *A New Theory Identifying the Locale of Columbus's Light, Landfall, and Landing.* Washington: Smithsonian Inst., 1964. 41 pp.

Smithsonian Miscellaneous Collections, v. 148, no. 1. Places CC's landfall on Watlings. The light was beyond, on High Cay, probably a fire kindled to drive off mosquitoes or other insects.

Martin Torodash

[310] **Roberto Barreiro-Meiro.** "Guanahani," *Rev Gen Marina* (Madrid), 161 (Dec 1966): 587-600.

Watlings is Guanahani; Cat is not.

Antonio Martín-Nieto Mora

[311] **Roberto Barreiro-Meiro.** "Guanahani de Ponce de León," *Rev Gen Marina* (Madrid), 173 (Oct 1967): 354-60.

Advances Herrera's account of the voyage of Ponce de León to confirm his previous conclusion (#310) that Watlings is Guanahani.

Antonio Martín-Nieto Mora

[312] **Antonio Rumeu de Armas.** "Descripción geográfica de la isla de Guanahani," *Anuario de Estudios Atlánticos,* no. 14 (1968), pp. 305-361.

Rpt. Madrid: Patronado de la "Casa de Colón," 1968. 57 pp. A methodical analysis and exposition of the geographical information on Guanahani and the surrounding Bahaman islands provided by CC's *Journal* and by parallel information in Las Casas' *Historia* and Ferdinand C's *Historie.* Intended to aid in determining which of the Bahamas is G.

[313] **Ramon J. Didier Burgos.** *Análisis del "Diario de Colón": Guanahani y Mayaguain; las primeras isletas descubiertas en el Nuevo Mundo.* Santo Domingo: Editorial Cultural Dominicana, 1974. 424 pp.

Index of names and places. An elaborate argument placing the landfall on the western of the two Plana Cays, northwest of Turks and Caicos Islands and southeast of Samaná Cay and Watlings Island. Includes reprint of most of CC's *Journal.*

[314] **Arne B. Molander.** "Columbus Landed Here—Or Did He?" *Américas* (Wash., DC), 33 (Oct. 1981): 3-7.

Guanahani is Egg Island, off the north cape of Eleuthera.

John Parker

[315] **John Parker.** "The Columbus Landfall Problem: A Historical Perspective," *Terr Incog*, 15 (1983): 1-28.

Rpt. *In the Wake of Columbus* (#19), pp. 1-28. Reviews the successive attempts to identify San Salvador, including the proposals of Mark Catesby, ca. 1747 (Cat Island); Juan Bautista Muñoz, 1793 (Watlings); Navarrete, 1825 (Turks); Samuel Kettell, 1827 (Grand Turk); Washington Irving and A.S. Mackenzie, 1828 (Cat); Alexander Von Humboldt, 1837 (Cat); George Gibbs, 1846 (Grand Turk): A.B. Becher, 1856 (Watlings); R. H. Major, 1870 (Watlings); G. V. Fox, 1882 (Samaná); J.B. Murdock, 1870 (Watlings); C.R. Markham, 1894 (Watlings); J.W. McElroy, 1941 (Watlings); S.E. Morison, 1942 (Watlings); R.T. Gould, 1945 (Concepción Cay); Pieter Verhoog, 1947 (South Caicos); E. & M. Link, 1958 (So. Caicos); R.J. Didiez Burgos, 1974 (Plana Cays); and Arne Molander, 1981 (Egg Island, off Eleuthera).

[316] **Oliver Dunn.** "Columbus's First Landing: The Evidence of the Journal," *Terr Incog*, 15 (1983): 33-50.

Rpt. *In the Wake of Columbus*, pp. 33-50. Probes, with the aid of a new transcription of CC's Journal (#55), Morison's 3 arguments for Watlings, i.e., (1) that a plot of CC's crossing points to Watlings; (2) that Watlings alone fits CC's description of San Salvador; (3) that only Watlings, as a first landing site, fits a backward tracing of CC's route through the islands from SS to Bahia Bariay in Cuba. Dunn argues (1) that the plot used by Morison (prepared by John W. McElroy; v. #666) involves indeterminate assumptions and a circular argument; (2) that although Watlings fits CC's description fairly well, so do other Bahaman islands; (3) that Morison's interpretation of the *Journal's* description of CC's sailing course between SS and Cuba corresponds closely enough to the actual geography to suggest strongly but not prove (in view of several problems) that the first landfall was Watlings.

[317] **James E. Kelley, Jr.** "In the Wake of Columbus on a Portolan Chart," *Terr Incog*, 15 (1983): 77-111.

Rpt. *In the Wake of Columbus*, pp. 77-111. Analyzes the *Journal* data on the first voyage, employing modern mathematical and statistical methods but a portolan projection to avoid the different distortions encountered on later projections. Through a computer simulation emphasizing the effect of local currents, identifies Acklin-Crooked Island as Isabela, CC's third island after the landfall. Since this furnishes a strong presumption that the landfall was Watlings, concludes by asking whether the *Journal* data and local currents can be interpreted so differently as to place Isabela anywhere but in the vicinity of Acklin-Crooked island.

[318] **Arne B. Molander.** "A New Approach to the Columbus Landfall," *Terr Incog*, 15 (1983): 113-149.

Rpt. *In the Wake of Columbus*, pp. 113-149.

Uses three lines of argument for a landfall in the northern Bahamas at Egg Island off Eleuthera: (1) rather than dead reckoning CC used the same accurate latitude sailing as his contemporaries (including John Cabot) to arrive at the location (Egg Island) indicated by the sum of the distances and bearings recorded in the *Journal*; (2) the weight of the evidence from the 12-day cruise through the Bahamas gives overwhelming support to this northern route; (3) the only contemporary measurement of San Salvador's latitude (Ponce de Leon, 1513) coincides with Northern Eleuthera.

[319] **Robert H. Power.** "The Discovery of Columbus's Island Passage to Cuba, October 12-27, 1492," *Terr Incog*, 15 (1983): 151-72.

Rpt. *In the Wake of Columbus*, pp. 151-72.

On the basis of diagrams derived from the entries of the *Journal*, infers that Guanahani is Grand Turk; Santa María de la Concepción is at different times Provinciales (one of the Caicos) and the Caicos as a whole; Fernandina is Mayaguana, Plana Cays, and Acklin; Isabela is Great Inagua; the "isleo" next to Isabela is Little Inagua; and the Islas de Arena are the Ragged Islands.

[320] **Pieter Verhoog.** "Columbus Landed on Caicos," *Terr Incog*, 15 (1983): 29-34.

Rpt. *In the Wake of Columbus*, pp. 29-34.

A point-by-point restatement of the argument in #304.

[321] **Joseph Judge.** "Where Columbus Found the New World," *National Geographic* (Wash. DC), 170 (Oct. 1986): 566-99.

A lavishly illustrated article by Judge arguing enthusiastically, on the basis of the Control Data Corporation's computer-estimated plot of CC's log, that Samaná Cay is Guanahani. Includes extensive discussion of CC's route from Guanahani to Cuba.

[322] **Luis Marden.** "The First Landfall of Columbus," *National Geographic*, 170 (Nov 1986): 572-77.

Discusses the elements that entered into the computer plot of CC's 1st voyage that led Joseph Judge (#321) to conclude that Samaná Cay is Guanahani. Marden says, with more reserve, that until more convincing new data emerge, the location of Samaná Cay is the most probable position of CC's landfall.

[323] **Robert H. Brill, I. L. Barnes, S. S. C. Tong, E. C. Joel, and M. J. Murtaugh,** "Laboratory Studies of some European Artifacts Excavated on San Salvador Island," *Columbus and his World* (Ft. Lauderdale FL: CCFL, 1987), pp. 247-92.

Recent excavations at the Long Bay site uncovered artifacts of European manufacture intermingled with native Indian artifacts: glass beads, a coin, a metal buckle, etc. Lab studies have addressed the dating.

The coin: Castillian 1471-74. All the artifacts contain intentionally-added lead, with isotope ratios consistent with Spanish origins. The artifacts can be traced to 3 locations in Spain.

RHB, etc.

[324] **Robert H. Fuson.** "The Turks and Caicos Islands as Possible Landfall Sites for Columbus," *Columbus and his World* (Ft. Lauderdale FL: CCFL, 1987), pp. 173-84.

Argues for the Joseph Judge *National Geographic* 1986 identification of Samaná Cay as Guanahani. Cf. #321, as well as #s 306-308.

[325] **Donald T. Gerace.** "Additional Comments Relating Watlings Island to San Salvador," *Columbus and his World* (Ft. Lauderdale FL: CCFL, 1987), pp. 229-35.

An analysis of CC's *Journal* account of the landing at Guanahani and exploration of the island, based on the author's personal experience at Watlings and on interviews with 20th-century natives. Concludes that if Watlings is Guanahani, CC's landing was in Graham's Harbor, about at Bamboo Point, where a boat can be rowed 9 miles north and back in the time described in CC's *Journal*. The place CC cites as suitable for a fort is Cat Cay, an island at the end of North Point.

DTG

[326] **Charles A. Hoffman.** "Archaeological Investigations at the Long Bay Site, San Salvador, Bahamas," *Columbus and his World* (Ft. Lauderdale FL: CCFL, 1987), pp. 237-45.

Archaeological study at a site on Watlings Island close to the presumed landing-place of CC reveals prehistoric pottery in direct association with objects similar to those CC reported trading to the Indians. This evidence strongly supports the hypothesis that Watlings is Guanahani.

CAH

[327] **Arne B. Molander.** "Egg Island is the Landfall of Columbus—A Literal Interpretation of his Journal," *Columbus and his World* (Ft. Lauderdale FL: CCFL, 1987), pp. 141-71.

Amplifies M's argument in # 318. The CC *Letter to Santangel* places his discoveries at 26 deg. latitude (rather than the 24 of the central routes); 2nd, Ponce de León recorded the latitude of San Salvador as 25 " 40 ' (barely 10 miles north of Egg Island, which is off Eleuthera Island); 3rd, the 1537 Chaves "rutter" seems to place the landfall 6 miles southwest of Spanish Wells—on the south beach of Egg Island.

ABM

[328] **Mauricio Obregón.** "Columbus's First Landfall: San Salvador," *Columbus and his World* (Ft. Lauderdale FL: CCFL, 1987), pp. 185-95.

Attacks Joseph Judge's *National Geographic* article (#321) that claims to establish Samaná Cay as Guanahani.

[329] **Philip L. Richardson and Roger A. Goldsmith.** "The Columbus Landfall: Voyage Track Correction," *Oceanus* (Woods Hole MA), 30, no. 3 (Fall 1987), pp. 2-10.

Proposes further corrections for Luis Marden's application of wind and current corrections to CC's first voyage in "The First Landfall of Columbus," #322.

[330] **Paolo Emilio Taviani.** "Why We are Favorable for the Watling— San Salvador Landfall," *Columbus and his World* (Ft. Lauderdale FL: CCFL, 1987), pp. 197-212.

Rejects arguments for the Turks and Caicos Islands as the site of the Landfall, as well as those for Cat Island, Samaná, and Egg Island.

3. The Second Voyage

[331] **Agustín Muñoz Gómez.** "Los Jerezanos, y el segundo viaje de Cristóbal Colón," *Bol R Acad Hist* (Madrid), 12 (1888): 425-32.

The texts of two royal cedulas of Queen Isabel ordering payments to citizens of Jérez who furnished supplies for CC on his 2nd voyage, along with the records of payment.

[332] **Edward Gaylord Bourne.** "Columbus, Ramón Pané, and the Beginnings of American Anthropology," *Proceedings of the American Anthropological Society*, New series, 17 (1905-6): 310-348.

Presents Bourne's English translation of Pané's *Treatise of Friar Ramon on the Antiquities of the Indians* (310-38), and the Hakluyt edition of Richard Eden's translation of the *Epitome* of Pané's treatise that appears in Peter Martyr's first *Decade,* Bk. ix. Characterizes Pané's treatise, commissioned by CC, as "the first systematic study of American primitive customs, religion, and folklore."

[333] **Charles R. Eastman.** "The Reversus, a Fishing Tale of Christopher Columbus," *Americana* (NY), 11 (1916): 438-446.

Rpt. *Scientific Monthly* (N.Y.), 3 (1916): 31-40. Refers to accounts of Ferdinand C, Andrés Bernáldez, and Peter Martyr of the small fish used to catch turtles on CC's 2nd voyage. Although the incident is probably authentic, the accounts appear to confuse the *remora* or sucking fish with the *diodon* or porcupine fish.

Mug 28

[334] **Lucius Hubbard.** "Did Columbus Discover the Islands of Antigua and St. Martin?" *Geog Rev* (NY), 21 (1931): 584-94.

Concludes from the Chanca narrative and La Cosa's map that the islands given these names by CC on the second voyage were, respectively, Nevis and St. Kitts. He could not have seen the two in question, which were too far to the northeast.

[335] **Samuel Eliot Morison.** *The Second Voyage of Christopher Columbus from Cadiz to Hispaniola and the Discovery of the Lesser Antilles.* Oxford: Clarendon, 1939. 112 pp.

M narrates the material later used in *AOS,* II, 49-99, in conjuction with his own observations made while sailing the same route and examining the sites touched by CC. Two purposes: to trace CC's route among the Lesser Antilles and to establish, by return to early documents, what names CC gave to the various islands.

[336] **William Jerome Wilson.** "A Narrative of the Discovery of Venezuela (1494), in the Thacher Manuscript on Columbus and Early Portuguese Navigations," *Proceedings of the Third Convention of the Inter-American Bibliographical and Library Association* (New York: H. W. Wilson, 1941), pp. 279-300.

Describes the MS (in LC) and discusses its origin and history. Points out that in editing the parts on CC for the *Raccolta,* Guglielmo Berchet rendered *"third* navigation" as *"fifth,"* and consequently ascribed to CC's 3rd voyage (rather than his 2nd) the voyage of 5 caravels to search for pearls. Wilson suggests that CC's route on his 3rd voyage can be explained as a plan, not revealed to the Spanish monarchs, to reach the pearl fishery that the 5 caravels had found in 1494.

See C. E. Nowell's reservations, #339.

Mug 106

[337] **William Jerome Wilson.** "The Spanish Discovery of the South American Mainland," *Geog Rev* (NY) 31 (1941): 283-99.

Studies the geographic aspects of the problems raised by the discovery of the narrative allegedly sent to Venice by Angelo Trevisan in about 1502. This narrative tells how CC sent 5 caravels south from Hispaniola in 1494, which discovered Venezuela and the pearl fishery and made a circuit of the Caribbean. Includes an English translation of the Ms, pp. 284-88.

Speculates on the authorship of the document; the route of the alleged voyage; the support the narrative gives to the 1500 dating of the Juan de la Cosa world map; CC's own apparent use of the narrative; CC's own apparent use of the narrative; CC's 1498 report on the "gardens" in Venezuela; and on evidence that CC may have made a confusing interpolation in the 1498 report to conceal the 1494 discovery from the sovereigns. See C. E. Nowell's reservations, #339.

[338] **William Jerome Wilson.** "The Historicity of the 1494 Discovery of South America," *Hisp Amer Hist Rev*, 22 (1942): 193-205.

Assesses the validity of the Angelo Trevisan MS on folios 68-71 of the 1904 J.B. Thacher MS, an account of a 10 Oct 1494 discovery of South America commissioned by CC from Hispaniola. Concludes that the account fits into what is known of CC's activities and intentions in 1493-94, and thus must be seriously considered. See C. E. Nowell's reservations, #339.

[339] **Charles E. Nowell.** "Reservations Regarding the Historicity of the 1494 Discovery of South America," *Hisp Amer Hist Rev*, 22 (1942): 205-210.

Holds that Wison's arguments (see #s 336,337,338)—supporting the possibility of the 1494 discovery reported in the Angelo Trevisan account—stand alone without the support of any other document and defy both the testimony in the *Pleitos* of CC's own men and the probability that some other report of the discovery would have come from one or more of the scores of sailors who would have taken part in the 1494 expedition.

[340] **Pedro Catalá Roca.** "Los Monjes que acompañáran a Colón en el 2o viaje," *Studi Colombiani* (Genoa: SAGA, 1952), 2: 371-90.

Develops the argument that the monks CC took along with Fray Bernal Buyl were Benedictines from the Montserrat monastery in Catalonia.

[341] **Filiberto Ramírez Correa.** *Reconstrucción Crítica del segundo viaje cubano de Colón: La ficción colombina del Cura de los Palacios.* Havana: Archivo Histórico Pinero, 1955.

The south coast of Cuba does not fit the account of this part of the second voyage as told by Andrés Bernáldez.

Martin Torodash

[342] **Saul Jarcho.** "Jaundice during the Second Voyage of Columbus," *Revista de la Associación de Salud Pública de Puerto Rico* (San Juan), 2 (1959): 24-27.

A study of illness aboard the *Niña* during CC's 1496 return voyage.
SEM, *The European Discovery of America*, II.140, #251.

[343] **Guillermo Esteves Völckers.** *Tarjetero histórico: Noticias sobre el segundo viaje del Almirante d. Cristóbal Colón.* Madrid: Privately published, 1960. 656 pp.

This "tarjetero" or "card-file" assembles notes, records, maps, and comments relating to CC's 2nd voyage from the time of the voyage itself until 1959. Bibliography of sources.

In 1964 Esteves published an *Appendice al Tarjeto Histórico* (Madrid: Privately published), 431 pp. Additional records, notes, maps, and comments related especially to the locales of Aguada and Aguadilla in western Puerto Rico.

[344] **Roberto Barreiro Meiro.** "Bojeo de Puerto Rico por Colón," *Rev Gen Marina* (Madrid), 176 (1969): 423-532.

Rpt. Madrid: Instituto Histórico de la Marina, 1969. Rejects the tradition begun by Peter Martyr, in contradiction of all the other sources, that CC sailed south of PR on his way to his destination, Navidad, on this second (outward) voyage.

[345] **Juan Manzano Manzano.** *Colón descubrió América del Sur en 1494.* Caracas, 1972. 493 pp.

Analyzes the chroniclers Bernáldez, Gómara, Las Casas, Oviedo, and Peter Martyr, and the *Pleitos,* and concludes (1) that in late 1493 CC sent an expedition from Hispaniola that explored the

South American coast from Cumaná almost to the later site of Cartagena; (2) that in Dec 1494 CC himself sailed southwest and discovered Guyana, Trinidad, and the pearl coast, whose wealth of pearls he concealed from the sovereigns.

R. Ezquerra

[346] **Cayetano Coll Toste.** *Cristóbal Colón en Puerto Rico: Llegada de los conquistadores españoles a Borinquén.* Sharon CT: Troutman, 1972.

Reviews the pertinent early documents and concludes that CC's landfall in PR was at the locale of the town of Aguada at the western end of the island.

[347] **Roberto Barreiro Meiro.** "La aguada de Colón en Puerto Rico," *Rev Gen Marina* (Madrid), 187 (1974): 453-58.

The place where CC took on water on the 2nd voyage is the bay formed by the headlands of Borinquen and Jigüero. This is consonant with the author's contention in #344 that CC's route on the 2nd voyage was along the north coast, not a circumnavigation of the island.

Isabelo Macías Domínguez

[348] **Demetrio Ramos Pérez.** "Colón y el enfrentamiento de los caballeros: un serio problema del segundo viaje que nuevos documentos ponen al descubierto," *Rev Indias* (Madrid), 39 (1979): 9-87.

A new vision of the circumstances in the 2nd voyage, which reveals that before the Roldán rebellion a conflict arose between CC and the "lanzas jinetas," a group of officers placed in the expedition by the monarchs. These officers, led by Margarit, decided to return to Spain, where they lodged complaints against the C. brothers. Documentation from the Archivo General de Simancas. Cf. #350.

[349] **J. Brotons Pico.** "Un medico en el descubrimiento de América," *Rev Gen Marina* (Madrid), 199 (1980): 3-11.

Analyzes Dr. Alvarez Chanca's diary of the 2nd voyage for evidence of his scientific knowledge, his powers of observation, and his qualities as a narrator. Concludes that these qualities make the document a basic source of knowledge about the voyage.

Isabelo Macías Domínguez

[350] **Demetrio Ramos Pérez.** *El conflicto de las lanzas jinetas: el primo alzamiento en tierra americana, durante el segundo viaje colombino.* Valladolid: Casa-Museo de Colón, 1982. 187 pp.

Proposes that the conflict between the C. brothers and the colonists that precipitated the Roldán revolt resulted from two distinct ways of understanding the problem of authority in Hispaniola. In its first phase the anti-C group constituted 20 officers of the anti-noble Hermandad, placed in the expedition by the monarchs, who hoped thus to keep power in their own hands and out of the hands of any anti-monarchical group. Ramos identifies the 20 "lanzas jinetas" and traces the conflict from the initial landing in Hispaniola to the return of the group to Spain. Cf. # 348.

[351] **Demetrio Ramos Pérez.** *Las variaciones ideológicas en torno al descubrimiento de América: Pedro Martír de Anglería y su mentalidad.* Valladolid: Casa-Museo de Colón, 1982. 84 pp.

Cuadernos Colombinos no. 10. Studies the evolution of Peter Martyr's assumptions about ancient learning: his movement from the humanistic exaltation of the ancients, to rejection based on the emergence of a new continent and new peoples unimaginable to the ancients, to a reconciliation of the new reality with the older world view. Much attention to the writing of Ramon Pané as catalyst to the thought of Peter Martyr.

[352] **Consuelo Varela.** "Dr. Diego Alvarez Chanca, cronista del segundo viaje colombino," *Historiografía y Bibliografía Americanistas* (Seville), 29 (1985): 1-48.

Rpt., *Temas Colombinos* (Seville: Escuela de Estudios Hispano-Americanos, 1986), pp. 1-48. A study of Chanca's life and his chronicle: (1) the known facts, with titles and locations of the 7 official Spanish documents and 40 other private docs referring to Chanca, 1491-55; (2) a study of Chanca's narrative in comparison with those of Peter Martyr, Coma, Cuneo, Bardi, and Verde; (3) a study of Bernáldez's use of Chanca's narrative in his account of the 2nd voyage, including parallel passages and an account of Bernáldez's misreadings.

[353] **Elpidio José Ortega.** *La Isabela y la arqueología en la ruta de Colón.* San Pedro de Macoris, D.R.: Universidad Central del Este, 1988. 100 pp.

(1) Pp. 9-59. An account of the findings, including photos and maps, of an archaeological study of CC's route on his first trip inland from La Isabela beginning 12 Mar 1494. (2) Pp. 61-79. A proposal for an archaeological re-evaluation of the presumed site of Isabela. Includes photos, charts, and illustrations.

4. The Third Voyage

[354] **E. Castellar Ripoll.** "El tercer viaje de Cristóbal Colón," *El Centenario* (Madrid), 2 (1892), 337-51.

Recounts the gradually deteriorating situation of CC from the time of his return to Hispaniola in 1498 until his arrest by Bobadilla in 1500. Largely a defense of CC; unfortunately not annotated.

[355] **Lucius L. Hubbard.** "Did Columbus Discover Tobago?" *Essays offered to Herbert Putnam by his Colleagues and Friends on his 30th Anniversary as Librarian of Congress, 5 April 1929*, ed. Wm. W. Bishop and Andrew Keogh (New Haven: Yale Univ. Press, 1929), pp. 211-23.

Although Las Casas identifies the islands of Ascunción and Concepción, discovered by CC after he left the Gulf of Paria on his 3rd voyage, with Tobago and Concepción respectively, CC could not have seen Tobago and Concepción on his famous direct voyage to Hispaniola.

[356] **Leonard Olschki.** "Ponce de León's Fountain of Youth: the History of a Geographical Myth," *Hisp Amer Hist Rev*, 21 (1941), 361-85.

Pp. 378-81, sets Ponce's belief in the fountain in the context of CC's geographical conceptions, in which (as with most of his contemporaries) the boundaries between different aspects of knowledge were fluctuating and interchangeable. Olschki illustrates this elaborately by the interpenetration, in CC's account of the discovery of South America (1) of genuinely scientific observations about the continental character of the land and the existence of the equatorial current and 2) speculations about the Earthly Paradise, which derive jointly from theological authorities and from CC's deductions from the mildness of the temperature and the huge volume of the Orinoco flow.

[357] **R. P. Devas.** "Discovery of Tobago and Grenada," *The Bajan* (Bridgetown, Barbados), June 1947, pp. 26-27; July, pp. 18-20; Aug., pp. 17-30.

Questions positioning of CC's fleet in emerging from the Boca de la Sierpe near Trinidad. CC discovered both Tobago and Grenada on 13 Aug 1498. He had not seen Tobago previously on 4 Aug, as Morison holds in *AOS*, #132.

[358] **Arthur Davies.** "The 'Miraculous' Discovery of South America by Columbus," *Geog Rev* (NY), 44 (1954): 573-82.

Suggests that CC never intended to sail southwest from the Canaries on the 3rd voyage until he reached Madeira on the way to the Canaries, where he learned of the Portuguese discovery of Brazil and thus of SA, and also learned of the best way to sail for SA.

[359] **J. Marino Incháustegui Cabral.** "Entorno a uno de los mas trágicos episodios de la vida de Colón," *Anuario de Estudios Americanos* (Seville), 24 (1967): 839-49.

There were 3 or even 4 Francisco de Bobadillas appearing in Spanish documents at the time when one of them removed CC from his governorship of Hispaniola in 1500. Incháustegui, puzzled by Oviedo's designation of him as a "very honorable and religious man," discovered that this was the *Corregidor* Bobadilla, who died in 1496. The records show that the *Comendador* Francisco Bobadilla, who cashiered CC and sent him home in chains, was a violent and unpredictable character.

[360] **Louis-André Vigneras.** *The Discovery of South America and the Andalusian Voyages.* Chicago: Univ. of Chicago Press, 1976. 170 pp.

Begins with "The Discovery of South America, its Causes and Consequences," pp. 1-22, arguing that CC decided to sail southwest looking for *tierra firma* because John Cabot had recently discovered a continent in the northwest. When the sovereigns learned of CC's discovery of a continent and of his being kept from exploiting the new discovery by the Roldán rebellion, they commissioned numerous navigators to explore in CC's place.

The rest of the book gives the background and details of these voyages by Hojeda, La Cosa, and Vespucci; Niño and Cristóbal Guerra; Vicente Yañez Pinzón; de Lepe; Vélez de Mendoza and Luis Guerra; Bastidas and La Cosa; Hojeda, Vergara, and Campos; C. and L. Guerra, La Cosa, and Ledesma; and Hojeda and Cueva. Treats in detail the methods of financing explorations.

[361] **J. H. Parry.** *The Discovery of South America.* London: Elek, 1979. 320 pp.

Italian ed. *La Scoperta del Sudamerica* (Milano: Mondadori, 1981), 379 pp. Treats the voyages of discovery to the West Indies and Central America as well. Seeks to establish the general significance of the process of discovery.

3 sections: (1) "The Unexpected Continent," treating the voyages; (2) "The Indian Empires," treating the encounters with the Aztec, Mayan, and Peruvian cultures; (3) "The Empty Spaces," treating the exploration of the Amazon and Orinoco, the Plata, Patagonia, and the Strait of Magellan.

[362] **Juan Gil.** "El rol del tercer viaje colombino," *Historiografía y Bibliografía Americanistas* (Seville), 30 (1985): 83-110.

In the tradition of Alicia Bache Gould, #260, presents (from the *Libro de armadas* in the Archive of the Indies, Seville) a list of 226 persons in the ships of the 3rd voyage, with a careful analysis of the information and its defects.

Includes an account of the double division of the fleet: *Santa Clara* (i.e., the *Niña*) and *Santa Cruz* sailed from Sanlucar de Barrameda 6 Feb 1498; later the *Castilla, Rábida, Gorda, Garza,* and *Santa María de Guía* sailed from Seville under CC, and were joined 24 May in Sanlucar by the *Vaqueña*. At the Canaries, 3 of these went straight to Hispaniola and the others, under CC, sailed a southern route to Trinidad and South America. Cf. J.M. Martínez-Hidalgo, #717.

[363] **José Enrique López de Coca Castaner.** "Publicidad en torno al tercer viaje colombino: fragmento de una carta de Juan Claver a Ludovico el Moro (enero de 1499)" *Pres Ital Andalu II* (Bologna: Cappelli, 1986), pp. 233-42.

Prints as appendix a hitherto unpublished postscript to a letter written by Juan Claver to Ludovico Sforza, Duke of Milan. The postscript consists of a fantastic account of CC's 3rd voyage, including a fabulous mountain shaped like a pear or a woman's breast. The article itself provides as background an account of the Juan Claver letter. López speculates on the reasons for this exaggeration of the material in CC's own account, and attributes it to the general desire to astonish one's readers when writing about the newly discovered lands.

5. The Fourth Voyage

[364] **Césareo Fernández Duro.** "El estrecho que buscaba Colón," *El Centenario* (Madrid), 3 (1893): 72-84.

A speculation on CC's attempt to interpret the geographies he had read and apply them to find a strait through Central America on the 4th voyage.

[365] **E. G. Bourne.** "The Voyages of Columbus and John Cabot," *The Northmen, Columbus and Cabot* (New York: Scribner's, 1906), pp. 77-418.

See notes to pp. 391-92 for the chronology of the 4th voyage while along the Miskito Coast of Honduras.

[366] **León Fernández.** "El pueblo de Cariarí," *Colección de Documentos para la historia de Costa Rica* (San José: Libreria Atenea, 1952), pp. 299-301.

The island Quiribri or Quiribiri and the town of Cariarí, encountered by CC on Voyage 4 on 17 Sept 1502, are not in the territory of Nicaragua but in that of Costa Rica.

[367] **Consuelo Varela.** "El rol del cuarto viaje colombino," *Anuario de Estudios Americanos* (Seville), 42 (1985): 243-95.

In the tradition of Alicia Bache Gould (#260) details (1) the sources, (2) criteria for arranging the ships' rolls, (3) the roll of 146 crew members with whatever information is available for each, e.g., salary payments; (4) a list of 5 doubtful crew members; (5) the names, and owners' names where available, of the 4 ships, i.e., *Capitana, Santiago de Palos, Gallega, and Vizcaino;* (6) analytical comments on this information; and (7) 6 royal cedulas of 1504 and 1505 referring to the payments made to the crews.

F. COLUMBUS IN SPAIN, 1493 – EMBARCATION DATE 1502

[368] **Antonio Rumeu de Armas.** "Colón en Barcelona: las bulas de Alejandro VI y los problemas de la llamada exclusión aragonese," *Anuario de Estudios Americanos* (Seville), 1 (1944): 431-524.

Rpt. Seville: Ed. Católica, 1945. 88 pp. A study of the curious fact that CC's famous 1493 reception in Barcelona, reported by many chroniclers, is nowhere mentioned in the rich and detailed Aragonese archives. Concludes that though this reflects the exclusion of Aragón from the Alexandrine bulls granting rights to Castile in the newly discovered lands, the exclusion is more apparent than real, a diplomatic maneuver by Ferdinand to take advantage of the rights of Castile in the ocean—recognized by the papacy—which Aragón did not have. The maneuver was intended to pre-empt any claims Portugal might make to the new lands.

[369] **Marina Conti.** "Le postille di Cristoforo Colombo alla 'Naturalis Historia' di Plinio il Vecchio," *Temi Colombiani* (Genoa: ECIG, 1986), pp. 76-91.

Advances reasons for inferring that CC's notes in Pliny's *Natural History* were made while CC was in Spain in 1496-98.

G. FINAL YEARS, 7 NOV 1504 – 1506

[370] **V. M. Fernández de Castro.** "Lo que hay sobre la Casa de Colón en Valladolid," *Revista Literaria* (Valladolid), 10 Apr 1878.

Rpt. *Boletin de la Sociedad de la Geografía de Madrid*, 24 (1888): 31-42. The only written source of the tradition that CC owned and died in the house at 2, Calle Aneha de la Magdelena in Valladolid is an unsupported note to this effect in D. Matías Sangrador's *Historia de Valladolid*, I.309. Sangrador claims that the house has always been in the hands of CC's family, but Fernández asserts that in fact it was owned from 1581 by the family Rivadeneyra, and that there is no evidence that the house had been built yet when CC died in 1506. Concludes that there is no evidence at all on the subject.

[371] **R. Vázquez Illa.** "La Casa de Colón en Valladolid," *Boletin de la Sociedad Geográfica de Madrid* 24 (1888): 22-31.

Continues the inquiry begun 10 years before by V. Fernández de Castro, #370, whose article is reprinted immediately following this one by Vásquez. Conclusion: the present generation does not know which house in Valladolid was inhabited by CC.

[372] **Cesáreo Fernández Duro.** "Noticias de la muerte de d. Cristóbal Colón y del lugar de enterramiento en Valladolid," *Bol R Acad Hist* (Madrid), 24 (1894): 44-46.

Quotes 10 contemporary 16th-c documents attesting to CC's death in Valladolid in May 1506 and 5 attesting to the transfer of his body to Las Cuevas in Seville.

[373] **Piero Gribaudi.** "Navigatori, banchieri, e mercanti italiani nei documenti degli archivi di Siviglia (Sec. xvi)," *Bollettino della R. Società Geografica Italiana* (Rome), Ser. 7, 1 (1936): 13-22.

Includes a transcript of the 1509 order of Johan Rodríguez for the deposit of CC's body in the Las Cuevas monastery in Seville, at the request of Diego Columbus, who moved the body from Valladolid.

Martin Torodash

V

Columbiana

A. RELATIONS WITH MONARCHS AND THE FORMATION OF COLONIAL POLICY

[374] **Fidel Fita,** ed. "Fray Bernal Buyl y Cristóbal Colón: nueva colección de cartas reales, enriquecida con algunas inéditos," *Bol R Acad Hist* (Madrid), 18 (1891): 173-233.

45 letters from the Catholic monarchs to Buyl and other figures concerning the priest and his relationship with CC.

[375] **Angel de Altolaguirre.** "Estudio jurídico de las capitulaciones y privilegios de Cristóbal Colón," *Bol R Acad Hist* (Madrid), 38 (1901): 279-94.

Concludes that the *Capitulations of 1492* made CC viceroy only of those lands discovered on the 1st voyage, and that the monarchs retained the right to terminate his tenure of the office.

[376] **Clarence H. Haring.** "The Genesis of Royal Government in the Spanish Indies," *Hisp Amer Hist Rev,* 7 (1927): 141-91.

Evaluates CC's and Bobadilla's failure to govern effectively, and Ovando's success; discusses the Columbus lawsuit against the Spanish crown, 1508-36.

Mug 38

[377] **Andrés María Mateo.** *Colón e Isabel la Católica; ensayo crítico sobre el carácter, el estilo y la letra del almirante, a través de una carta suya autógrafa y mutilada a la reina.* Valladolid: Seminaria Simancas, 1942. 130 pp.

Publishes and analyzes a letter to Queen Isabel found in the Archive of Simancas. Concludes that although the signature and date are missing, the letter is in CC's hand, written about Aug-Sept. 1501. The letter reveals CC's tenacious, pious, melodramatic nature and his tendency to suggest and insinuate the points he hopes to put across.

[378] **Samuel Eliot Morison.** "The Earliest Colonial Policy toward America: that of Columbus," *Bulletin of the Pan Amer Union* (Wash, DC), 76 (1942): 543-550.

CC's ideas derive from his knowledge of the Portuguese experience along the African coast and in the offshore islands. CC wished especially to found a trading post near China—a wish carried out in 1571 with the establishment of Manila.

Martin Torodash

[379] **Florentino Pérez Embid.** "El Almirantazgo de Castilla, hasta las Capitulaciones de Santa Fe," *Anuario de Estudios Americanos* (Seville), 1 (1944): 1-170.

A study (1) of the office of Admiral from its obscure origins in the activity of Ramón Bonifaz, 13th-c founder of the Castilian fleet, and (2) of the persons who held the office. Intended as a basis for understanding the significance of CC's office as Admiral of the Ocean.

[380] **Juan Manzano Manzano.** *Por que se incorporaron las Indias a la corona de Castilla?* Madrid: Cultura Hispánica, 1948. 356 pp.

Studies (1) the arguments developed by Solórzano Pereira in the 17th c to justify the Spanish penetration into the New World, and (2) why the Indies were annexed to the crown of Castile, i.e., because this was the best expedient for avoiding difficulties with the troublesome nobles of Aragón.

[381] **Paolo Revelli.** "La concezione coloniale di Cristoforo Colombo," *Memorie della Accademia Nazionale dei Lincei* (Rome), Ser. 8, 1 (1948): 329-56.

CC's Letter of 4 Mar 1493 places the discoverer among the precursors of the modern colonial conception, having thought out and partly set in motion the exploitation of the newly discovered lands by means of native labor, selected Spanish settlers, and the construction of new settlements.

[382] **Andrés Angulo Pérez.** "La odisea colombina y el destino de una empresa," *Studi Colombiani* (Genoa: SAGA, 1952), 2: 357-61.

Infers the effects of the chance linkup between CC and the Spanish monarchs: the improvisatory character of the colonial government, the dominance of Castilian institutions in the American colonies, and the tragic effects of the anachronistic application of medieval institutions in America, such as the use of the Inquisition to justify outrages to the aborigines.

[383] **Charles Verlinden.** "Colomb et les influences médiévales dans la colonisation de l'Amérique," *Studi Colombiani* (Genoa: SAGA, 1952), 2: 407-418.

Points out the heavily medieval and feudal cast of the privileges won by CC in the *Capitulations* and the strong tendency of CC's viceroyalty (as inherited by Diego Colón) to continue this. The monarchs' increasing tendency to limit these privileges is consonant with their tendency to centralize authority, and represents the evolution of the medieval tradition. CC as a Genoese reflects the central part played by Genoa in the economic and colonial policy of Spain and Portugal, a signal example of evolution of medieval practices. Concludes with a call for organized study of Genoese influence on Spanish and Portuguese colonial development, anticipating (and probably inspiring) Jacques Heers' study *Christophe Colomb* (#137).

[384] **Manuel Ballesteros Gaibrois.** "'Fernando el Católico' y América," *Anales de la Universitá de Santo Domingo* (DR), 9 (1954): 353-69.

Reviews a series of actions by King Ferdinand indicating that he was not unsympathetic to CC or determined to undermine CC's claim to the privileges granted him in 1492.

[385] **Juan Pérez de Tudela.** "La negociación colombina de las Indias," *Rev Indias* (Madrid), 14 (1954): 289-357.

Proposes a full restudy of the Spanish colonization begun by CC, to 1505 (see items 387 through 389 for the continuation of this restudy). Anticipates a thorough classification and scrutiny of the documents. The central question is: what were CC's ideas for systematically directing the settlement and exploitation of the lands he discovered?

The sovereigns displayed a concerted and unified program in their activities generally, and the conflict between faith and acquisitiveness had not yet developed. CC, while undoubtedly sincere in his expressed desire to convert the Indians, brought to his activity a peculiarly Genoese element that controlled his Enterprise: i.e., his policy of *negotiating* to best advantage.

Analyzes the elements of CC's ambitious attempt to exploit Hispaniola up to 1496 when he returned to Spain to defend himself against the accounts brought back by disillusioned colonists.

[386] **Sigfrido A. Radaelli.** "La institución virreinal en las Indias. Antecedentes históricos," *Rev Indias,* 14 (1954): 37-56.

The office was discontinuous, not evolutionary. Finds no precedent in Castile for CC's office of viceroy. Later American viceroyalties, after 1535, were also independent of precedents, including that of CC's office itself. Cf. #422.

Martin Torodash

[387] **Juan Pérez de Tudela.** "Castilla ante los comienzos de la colonización de las Indias," *Rev. Indias* (Madrid), 15 (1955): 11-88.

Continues #385. Whereas CC came from a mercantile tradition of exploiting colonies, the Castillian tradition was repopulation of the lands taken under the Spanish flag, e.g., the Moorish lands in Spain and the lands taken from the Canary natives. This study explores the response to CC's policy of exploitation and settlement through negotiation under a centralized viceroyalty, which collapsed and was replaced gradually by parallel explorations authorized by the crown and further efforts at colonization. See #s 388 and 389.

[388] **Juan Pérez de Tudela.** "La quiebra de la factoría y el nuevo poblamiento de la Española," *Rev Indias* (Madrid), 15 (1955): 197-252.

Continues 385 and 387, tracing the collapse of CC's viceroyalty as Bartholomew C, and then CC himself on his return from Spain, grappled unsuccessfully with the Roldán rebellion. The sovereigns meanwhile authorized new explorations that were bound to lead to new colonies. Ironically, Roldán engaged in negotiation with the natives that ultimately led to stabilization. See #389.

[389] **Juan Pérez de Tudela.** "Política de población y política de contratación de las Indias (1502-1505)," *Rev Indias* (Madrid), 15 (1955): 371-420.

Continues nos. 385, 387, and 388. Traces the working out of the situation in the new Spanish colonies as the newer colonists accepted and articulated the combination of "repopulation" and mercantilism that had crystallized under Ovando. A prime and decisive element in stabilization was the conjugal relationship between Spaniards and the women of the leading Indian families.

[390] **Antonio Rumeu de Armas.** *Itinerarios de los Reyes Católicos.* Madrid: 1974.

A day-by-day, year-by-year record of the whereabouts of Ferdinand and Isabel from 11 Dec 1474, when Henry IV of Castile died, until 23 Jan 1516, when Ferdinand died. Whenever the monarchs are not together their whereabouts are recorded in separate columns. At the end of each triennial period, beginning 1474-77, appears a summary narrative and a map.

The documentary sources for each location are recorded in a separate column. Elaborate toponomic-chronological index. A work of fundamental importance for studying the relationship betwen the monarchs and CC, 1485-1606.

[391] **Francisco Morales Padrón.** *Teoría y leyes de la conquista.* Madrid: Cultura Hispánica, 1979. 536 pp.

Publishes, in 20 chapters, the papal bulls and royal Spanish documents that register the gradual evolution of Spanish policy toward the possessions in the New World. A discussion of the historical context and a bibliography accompany each chapter.

[392] **Demetrio Ramos Pérez.** *Audacia, negocios y política en los viajes españoles de "Descubrimiento y rescate."* Valladolid: Casa-Museo de Colón, 1981. 626 pp.

A highly detailed study, in 18 chapters, of Spanish voyages to and from the New World, 1495-1573. 40 docs appended, including relevant capitulations, asientos, letters, and cedulas.

Consuelo Varela

[393] **Rafael Diego Fernández.** *Capitulaciones Colombinos.* Zamora, Mexico: Colegio de Michoacán, 1987. 434 pp.

Studies 14 "capitulations," agreements between the Spanish crown and explorers sent to possess and exploit the Indies, beginning with CC. Attempts to define what was meant and understood by the term "capitulation."

B. THE ENTERPRISE OF THE INDIES

1. Background

[394] **Pinheiro Chagas.** "Las Novelas de los descubrimientos precolombinos," *El Centenario* (Madrid), 2 (1892): 330-36.

A common-sense debunking, prior to the claims of García de la Riega and many others, of all the prediscovery claims and other wishful claims that are based on a nationalistic or simply egoistic desire to discredit the importance and greatness of CC's voyage.

[395] **H. Yule Oldham.** "A Pre-Columbian Discovery of America," *Royal Geographical Society Journal,* (1895): 221-33.

Infers the discovery from the fact that the Bianco world map of 1448 (in the Biblioteca Ambrosiana of Milan, no. XI in the Ongania Collection) shows land to the west of Africa, "1500 miles distant" according to the usual intepretation of a legend close to the land.

[396] **B. F. de Costa.** *The Pre-Columbian Discovery of America by the Northmen with Translations from the Icelandic Sagas.* 1st ed. 1868. 3rd ed., rev., Albany NY: Munsell, 1901.

Chiefly an anthology of translations from the Icelandic sagas, including narratives of Gunnbjorn, Eric the Red (Greenland), Biarne Heriulfsson (American coast), Leif Ericson (Vinland), Thorston Ericson, Thorfinn Karlsefne (Vinland settlement). Also minor narratives, geographical descriptions, and English translations of a series of papal letters chiefly to archbishops of Drontheim, Norway, respecting circumstances in the churches of Greenland, A.D. 1206-1493.

[397] **Julius E. Olson and Edward Gaylord Bourne,** eds. *The Northmen, Columbus, and Cabot.* New York: Scribner's, 1903. 443 pp.

English translations. Under "The Northmen," Olson presents the Vinland narrations in the *Saga of Eric the Red* and in the *Flat Island Book* and extracts from Adam of Bremen and from the *Icelandic Annals,* plus relevant letters of Popes Nicholas V (1448) and Alexander VI (1492). Under "Columbus," Bourne presents the *Capitulations of 1492* and the royal letter granting CC's title (30 Apr 1492), Markham's version of the *Journal,* the *Letter to Santangel,* Chanca's *Letter* on the 2nd voyage, Las Casas' narrative of the 3rd voyage, and CC's letter on the 4th voyage and other matters. Under "Cabot" Bourne presents the letters of Pasqualigo and Soncino (1497) and Pedro de Ayala's letter to the Catholic Sovereigns of 25 Jul 1498.

C. R. Benzley

[398] **Jaime Cortesão.** "Do sigilo nacional sobre os descobrimentos," *Lusitania* (Lisbon), 1 (1924): 45-81.

The rulers of Portugal by policy destroyed or concealed maps and records of African discoveries, to impede interference from other nations. This would be a significant matter if the Portuguese reached America before CC.

C. E. Nowell

[399] **Sofus Larsen.** *The Discovery of North America Twenty Years before Columbus.* London: Hachette, 1924. 116 pp.

Cites apparent Portuguese-Danish cooperation that resulted in a Danish voyage to North America, ca. 1472-74, with Portuguese observers Corte-Real and Alvaro Martins Homem on board. Infers an exploration of the mouth of the St. Lawrence, Newfoundland, and Cape Breton.

C. E. Nowell

[400] **Jordão de Freitas.** "O descobrimento pre-Colombino da America Austral pelos Portugueses," *Lusitania* (Lisbon), 9 (1926): 315-27.

King John II's insistence on a line of demarcation 370 leagues west of the Cape Verde Islands implies that John was aware of the location of the Brazilian coastline.

C. E. Nowell

[401] **Jaime Cortesão.** "Le traité de Tordesillas et la découverte de L'Amérique," *Atti XXII Congresso Internazionale degli Americanisti,* 1926 (Rome, 1928), 2: 649-83.

Studies physical, scientific, and economic conditions related to the treaty that might have influenced the discovery of America. Argues that the treaty (1494) implied a pre-CC discovery in that King John's insistence on a line of demarcation 370 leagues west of the Cape Verde Islands makes sense only if he already knew of the eastward-thrusting Brazilian coastline. Cf. #407.

C. E. Nowell

[402] **Giuseppe Caraci.** "Una pretesa scoperta dell'America vent'anni innanzi Colombo," *Boll Soc Geog Ital* (Rome), 67 (1930): 771-812.

Analyzes and rejects Sofus Larsen's case (#399) for a Danish discovery of North America. 1472-74.

C. E. Nowell

[403] **Alberto Magnaghi.** *Precursori di Colombo: Il tentativo di viaggio transoceanico dei genovese fratelli Vivaldi nel 1291* Rome: Anonima Arte Grafiche, 1935. 155 pp.

Vigorously documented review of what is to be known about the Vivaldi brothers' ill-fated voyage. Shows that in 1291 Ugolino and Vadino Vivaldi planned to sail to India along the parallel of the Canaries, anticipating CC's plan by 2 centuries. The expedition wrecked off Morocco.

[404] **Egmont Zechlin.** "Das Problem der vorcolumbischen Entdeckung Amerikas" *Historische Zeitschrift* (München), 152 (1935): 1-47.

Summarizes the various claims for pre-Columbian discoveries, and seems favorable to the Portuguese claim.

C. E. Nowell

[405] **R. Hennig.** "Die These einer vorcolumbischen portugiesischen Geheimkenntnis von Amerika," *Historische Vierteljahresschrift* (Leipzig), 30 (1936): 548-92.

Rejects the Portuguese claim to a pre-Columbian discovery.

C. E. Nowell

[406] **Roger Bigelow Merriman.** *The Rise of the Spanish Empire in the Old World and the New.* New York: Macmillan, 1936. 2 vols.

V. 2, ch. 16 on the Canaries. Cites the occupation as a microcosm of the Spanish experience in the New World.

[407] **Jaime Cortesão.** "The Pre-Columbian Discovery of America," *Geog Jour* (London), 89 (1937): 29-42.

After claiming that transatlantic voyaging was sometimes easier than coastal, and without absolutely claiming a Portuguese pre-discovery of Newfoundland, rehearses what he sees as the many intimations in CC's writings, in the *Pleitos,* and elsewhere that the Portugese and others thought there was continental land west or southwest of the Azores; and points out the advantages to Portugal of adroitly concealing whatever knowledge it may have had, in order not to encourage Spanish navigations that might pre-empt the House of Avis's plan to reach the Indies around Africa.

Cortesão also contends that John II's insistence on pushing the Tordesillas treaty line to the west was due to the king's desire to preserve the southwest continent for Portugal. Cf. #401.

[408] **G. R. Crone.** "The Alleged Pre-Columbian Discovery of America," *Geog Jour* (London), 89 (1937): 455-60.

Reply to Jaime Cortesão's article, #407. To the claim that transatlantic voyaging could be easier than coastal, Crone replies that this oversimplifes the winds and currents and ignores the effect of the unknown. Respecting the official Portuguese secretiveness about its geographical knowledge, Crone says that it didn't work. Rejects the quasi-claim of pre-Columbian Newfoundland discovery as merely speculative, and rejects the idea that John II's pushing the Tordesillas line westward had any purpose except to protect the Portuguese route to the Indies.

[409] **Antonio Ballesteros Beretta.** *Génesis del descubrimiento.* Barcelona: Salvat, 1947. 766pp.

Vol III, *Historia de America.* A very thorough review of what was known in 1947 about the context within which CC's plan was generated. Examines ancient geographers; ancient historians; geographical and historical writings of the church fathers; geographical and historical writings of the Arabs; myths such as those of Prester John and John de Mandeville, and the partly factual narrative of Marco Polo; the evidence for pre-Columbian exploration by the Chinese and the Vikings; fantasies about the Atlantic, and the conquest of the Canaries.

A supplementary section of this book, by Jaime Cortesão, pp. 497-758, reviews the Portuguese ferment of discovery, including voyages southwest into the Atlantic which gave rise to the conjectures of "Portuguese prediscovery."

[410] **Florentino Pérez Embid.** *Los descubrimientos en el Atlantico y la rivalidad castellana-portuguesa hasta el tratado de Tordesillas.* Seville: Escuela de Estudios Hispano-Americanos, 1948. 370 pp.

A detailed survey of discoveries in the Atlantic, including these steps: (1) preliminary voyages to Atlantic islands beginning 1295; (2) fundamental developments: (a) the chief discoveries and the growth of Portuguese-Castilian rivalry (1415-94); (b) expansion into the newly opened areas (1494-1550); (c) English and Dutch naval dominance (1550-1750); (d) scientific explorations (1750-1800).

[411] **Armando Cortesão.** *The Nautical Chart of 1424 and the Early Discovery and Cartographical Representation of America.* Coimbra: Univ. of Coimbra, 1954.

Threefold purpose: to describe a Venetian nautical chart of 1424 of the Bibliotheca Phillipica, attributed to Zuane Pizigano, which includes an archipelago west of the Azores labeled "Antilia"; to show that it is the first document in which the word "Antilia" occurs; and to show that it represents and confirms an early Portuguese discovery of the West Indies and perhaps Florida.

[412] **David B. Quinn.** "The Argument for the English Discovery of America between 1480 and 1494," *Geog Jour* (London), 127 (1961): 277-85.

Constructs a scenario to explain a sentence in John Day's letter, presumably to CC, about John Cabot's voyage. The sentence in question seems to say that Bristol sailors discovered the mainland behind Newfoundland "in other times."

[413] **Antonio Ballesteros Beretta.** *La Marina Cántabra.* Santander: Diputación Provincial, 1968.

Vol. 1 of 4 vol. history, the remaining volumes by other hands. Within this history of the transfer of the traditions, skills, and personnel of the Cantabrian marine activity to the Andalusian coast, Ballesteros includes a significant account of Juan de la Cosa, whom he deems a native of Cantábria transferred to the Puerto de Santa María in Andalusia.

B. acknowledges the existence of a second J de la C on the second voyage, but argues convincingly that the owner and master of the *Santa María* of the 1st voyage was the mapmaker of the 2nd voyage, forced by CC with others to swear that Cuba was an island. This J de la C was also the maker of the 1500 mappamundi, which is correctly dated. The representation of Cuba as an island in this map probably expresses the mapmaker's true opinion and also his resentment at being forced to swear otherwise. Treatment of J de la C ends with acct. of his relations with Rodrigo de Bastidas and the Spinolas, and his further explorations in the Spanish Main.

[414] **Wilcomb E. Washburn,** ed. *Proceedings of the Vinland Map Conference.* Chicago: Univ. of Chicago Press, 1971. 185 pp.

Preface by Washburn, pp. ix-xvii, characterizes and comments on the 16 essays and the edited record of the general discussion which followed their presentation at the conference. The essays: Laurence Witten, "Vinland's Saga Recalled"; Armando Cortesão, "Is the Vinland Map Genuine?"; John Parker, "Authenticity and Provenance"; Robert S. López, "The Case is not Settled"; Thomas E. Goldstein, "Some Reflections on the Origin of the Vinland Map"; Melvin H. Jackson, "The Vinland Map and the Imperatives of Medieval Form"; Ib Ronne Kejlbo, "Claudius Clavus and the Sources of the Vinland Map"; Paul Fenimore Cooper, Jr., "The Representation of Greenland on the Vinland Map"; Boleslaw B. Szczesniak, "The Tartar Relation and the Vinland Map: Their Significance and Character"; Vsevolod Slessarev, "The Great Sea of the Tartars and the Adjacent Islands"; Konstantin Reichardt, "Linguistic Observations on the Captions of the Vinland Map"; Stephan Kuttner, "Observations on the Relationship between Church History and the Vinland Map"; Gwyn Jones, "The Western Voyages and the Vinland Map"; Erik Wahlgren, "The Companions Bjarni and Leif"; Einar Haugen, "Bishop Eric and the Vinland Map"; and Oystein Ore, "A Hypothesis for the Vinland Map."

[415] **Demetrio Ramos Pérez.** *Los contactos transatlánticos decisivos, come precedentes del viaje de Colón.* Valladolid: Casa-Museo de Colón, 1972. 66 pp.

Cuadernos Colombinos no. 2. Studies Atlantic voyages beginning in 1450, including those by the Portuguese. Addresses the question of the unknown pilot.

[416] **Louis-André Vigneras.** "La búsqueda del paraíso y las islas legendarias," *Anuario Estud Amer* (Seville), 30 (1973): 809-863.

Rpt., Cuadernos Colombinos no. 6. Valladolid: Casa Museo de Colón, 1976. 71 pp. The history and legend of the Atlantic islands in ancient and medieval times, with special reference to travelers' tales, e.g., St. Brendan and the monks of St. Matthew. Indicates the supposed locations of the islands and concludes with accounts of the main Portuguese and English voyages seeking the mythical isles.

Francisco Castillo Meléndez

[417] **Alberto Boscolo.** "Genova e Spagna nei secoli xiv e xv: una nota sugli insediamente," *Atti del Convegno Internazionale di Studi Colombiani* (Genoa: Civico Istituto Colombiano, 1974), pp. 37-49.

A documented survey of locales where Spanish and Genoese had established themselves. Touches on Aragonese establishments in Sicily and Sardinia and Genoese establishments in Aragón, Granada, and Castile, esp. in Córdova and Seville.

[418] **Jaime Cortesão.** "El viaje de Diogo de Teive," *El viaje de Diogo de Teive. Colón y los Portugueses.* (Valladolid: Casa-Museo de Colón, 1975), pp. 9-29.

Cuadernos Colombinos no. 5. Sets forth the available evidence, admittedly not conclusive, of a voyage by Teive and Pero Vásquez de la Frontera to the Newfoundland bank in 1452.

[419] **Thomas R. Adams.** "Some Bibliographical Observations on and Questions about the Relationship between the Discovery of America and the Invention of Printing," *First Images of America* (Berkeley: Univ. Cal Press, 1976), 2: 529-36.

The reason why little has ever been done to define the relationship lies (1) in the state of bibliography as a whole and (2) in the state of bibliographical control over "Americana." Reviews the history of the bibliography of Americana and suggests what remains to be done.

[420] **Robert L. Benson.** "Medieval Canonistic Origins of the Debate on the Lawfulness of the Spanish Conquest," *First Images of America* (Berkeley: Univ. Cal Press, 1976), 1: 327-34.

Cites Vitoria's *De Indiis prior* to illustrate his contentions (1) that the 12th and 13th-c rebirth of Roman and canon law created the preconditions for international law, and (2) that the problems which the Indians posed for the Spaniards were anticipated by the Crusaders' prior encounter with peoples who differed in language, culture, and religion.

[421] **Thomas Goldstein.** "Impulses of Italian Renaissance Culture behind the Age of Discoveries," *First Images of America* (Berkeley: Univ. Cal Press, 1976), 1: 27-35.

"The ultimate inner link between the Renaissance and discoveries is implicit in . . . [a new] relationship towards space and time." "It is not the mere existence of unknown lands . . . nor even the theoretical recognition of their existence, but the irresistible impulse to tear these lands from obscurity at just that particular moment which defines the Age of Discoveries as a historic turning point" (p. 32).

[422] **Charles E. Nowell.** "Old World Origins of the Spanish-American Viceregal System," *First Images of America* (Berkeley: Univ. Cal Press, 1976), 1: 221-230.

The Spanish American institution of viceroy, whose holder exerted authority not only in the monarch's name but as his *alter ego* theoretically identical with the monarch, originated in the Venetian-controlled Constantinople in the early 13th century and was employed in Mediterranean possessions by Genoa, Pisa, and Amalfi. It was adopted by Catalonia and Aragón for various units of their Mediterranean holdings in the 13th through 15th centuries. The Portuguese adopted the office for Almedia in 1505. The first Spanish viceroy after CC (whose viceroyalty quickly failed) was Antonio de Mendoza, viceroy of Mexico beginning in 1535. The office, used also in Peru, Colombia and Rio de la Plata, expired in 1824 with the defeat of the Peruvian viceroy José de la Sema by the Colombians. Cf. #386.

[423] **Raquel Soeiro de Brito.** "Les îles de l'Atlantique et leur rôle dans l'histoire des découvertes maritimes," *Atti II Conv Internaz Stud Col* (Genoa: CIC, 1977), pp. 125-57.

Before the 15th-c Iberian occupation of the Atlantic islands, the Atlantic was a hostile and uncivilized ocean. The communication of European culture to the islands transformed the ocean into an incipient Mediterranean. This served as a laboratory for the Europeanization of the Western Hemisphere and perhaps made possible the successful exploitations of the 16th c.

2. "El Secreto": The Unknown Pilot and Other "Prediscovery" Conjectures

[424] **Cesáreo Fernández Duro.** "La tradición de Alonso Sánchez de Huelva, descubridor de tierras incognitas," *Bol R Acad Hist* (Madrid), 20 (1892): 33-53.

The case for Sánchez as the unknown pilot on whom CC depended for his certainty about the location of the Indies.

[425] **José A. Aboal Amaro.** *El piloto desconocido: Un Andaluz descubrió América en 1484? Antología y bibliografía críticas.* Montevideo: Biblioteca Colombina, 1957. 111 pp.

Reviews 108 published opinions on the validity of the "unknown pilot" story, abroad since ca. 1502. 46 accept the story, 35 reject it, and 27 are on the fence. Bibliography, pp. 93-110.

J. A. Aboal Amaro

[426] **Juan Manzano Manzano.** *Colón y su secreto.* Madrid: Cultura Hispánica, 1976. 743 pp.

2nd ed., 1982. Develops, probably to its limit, the inference of a "pre-discovery" (that CC knew about or participated in) in the reference to "lo que ha descubierto" in the *Capitulations of Santa Fe* of 17 Apr 1492. Also develops the argument that CC discovered South America in 1494. An extremely detailed examination of the Enterprise of the Indies, including all 4 voyages.

[427] **Juan Manzano Manzano.** "Fray Antonio de Marchena, principal depositario del gran secreto colombino," *Andalucía y América en el Siglo XVI. Actas de las II Jornadas de Andalucía y América* (Seville: Escuela de Estudios Hispano-Americanos, 1983), 2: 501-516.

Analyzes what is known of Marchena up to the time CC got his commission. Provides support for M's contention throughout his career that Marchena was, with Diego de Deza, one of the two persons to whom CC entrusted his "secret."

[428] **Juan Pérez de Tudela.** *Mirabilis in Altis: Estudio crítico sobre el origen y significado del proyecto descubridor de Cristóbal Colón.* Madrid: Consejo Superior de Investigaciones Científicas, 1983. 429 pp.

Affirms as the origin of CC's first voyage the preknowledge that the discoverer gained about the Indian lands through information gathered, while CC was aboard a Portuguese vessel, from one or more West Indian women who had navigated to the middle of the Atlantic in their canoes. Concludes that CC's certainty about his geographic goal and his conviction about his prophetic mission stem from this decisive mid-Atlantic experience. Touches on CC's postils, the Earthly Paradise, the Amazonian culture, heroic femininity in relation to the cult of the Spanish queen, and the union of seeming contradictions in CC's nature.

[429] **Gabriella Moretti.** "Nec sit terris ultima Thule. (La profezia di Seneca sulla scoperta nel Nuovo Mondo)," *Columbeis I* (Genoa: IFCM, 1986), pp. 95-106.

Speculation on CC's inclusion of the famous verses from Seneca's *Medea* in his *Book of Prophecies*, including the observation that CC's translation (probably unconsciously) makes the prefiguration of the Enterprise of the Indies stronger.

[430] **Osvaldo Baldacci.** "Il segreto di Colombo: solo le rotte atlantiche del primo viaggio?" *Temi Colombiani* (Genoa: ECIG, 1987), pp. 13-29.

The story of CC's "secret" route to the Indies, though inspired by Iberian pride, assumes CC's habitual secretiveness about the route that he in fact planned methodically. His own assumptions included the aid of the trade winds and "seeking the latitude" of the terminal port. His computation of distances traveled seems to have depended on the "taoleta de marteloio" which converted sailing distances to progress along an east-west route.

[431] **John Larner.** "The Certainty of Columbus: Some Recent Studies," *History* (London), 73, no. 237 (Feb 1988): 3-23.

Reviews recent studies addressing the source of CC's serene confidence that he could reach the Indies, with emphasis on Manzano's *Colón y su secreto*, #426, and Pérez de Tudela's *Mirabilis in altis*, #482, both concerned with secret but mundane information CC had somehow received, and with attention also to the studies of CC's mystical and prophetic certitude in Alain Milhou's *Colón y su mentalidad mesiánica*, #637, and Pauline Moffitt Watts's "Prophecy and Discovery," #638.

3. Other Studies

[432] **Angel de Altolaguirre.** *Cristóbal Colón y Pablo del Pozzo Toscanelli: estudio crítico del proyecto, formulado por Toscanelli y seguido por Colón, para arribar al extremo oriente de Asia navegando la via del oeste.* Madrid: Administración Militar, 1903. 429 pp.

CC somehow came into possession of Toscanelli's plan, and proposed it first to John II of Portugal and then to Isabel and Ferdinand. Ferdinand Columbus, learning that T's plan was already known in Portugal when CC got there, invented and forged CC's correspondence with Toscanelli to cover up the fact that CC did not invent the Enterprise himself.

[433] **Henry Vignaud.** *Histoire critique de la grande entreprise de Christophe Colomb.* Paris: Welter, 1911. 2 vols.

Assembles and discusses with great thoroughness the facts of CC's life through the first voyage, to the degree that these were knowable at the time of publication. 1st vol. 1476-90; 2nd vol. 1491-93. Annotated list of documents.

[434] **Henry Vignaud.** *The Columbian Tradition on the Discovery of America and of the Part Played Therein by the Astronomer Toscanelli.* Oxford: Clarendon, 1920. 62 pp.

Sums up V's lifetime polemic to establish these points: (1) The 1492 expedition did not have the East Indies as its goal. (2) It had as its sole object the discovery of new lands. (3) CC had information about what he set out to find. (4) The Toscanelli map and correspondence are spurious.

[435] **Charles E. Nowell.** "The Discovery of Brazil—Accidental or Intentional?" *Hisp Amer Hist Rev* (Durham NC), 16 (1936): 311-38.

Pp. 317-320, reviews the evidence suggesting that CC's Enterprise of the Indies was formed while he was in Portugal. Includes a survey of CC's career in the Portuguese years.

[436] **Diego Luis Molinari.** *La Empresa Colombina.* Buenos Aires: Impr. de la Universidad, 1938. 230 pp.

Also in *Historia de la Nación Argentina* (Buenos Aires, 1939), 2: 341-528. 3rd ed. (1961), 2: 223-346. Reviews, in context, the genesis and the execution of the Enterprise of the Indies, to the death of CC. Morison praises this synthesis in *AOS* as the best available (#132).

[437] **Emiliano Jos.** "La Genesis Colombina del Descubrimiento," *Rev Hist Amer* (Mexico City), no. 14 (1942), pp. 1-48.

Surveys the formation of CC's Enterprise of the Indies, touching on the root problems of geography and the scientific context; on the participation of Bartholomew C; on the influence of CC's Genovesità and his presence in Lisbon and Galway; on CC's cosmographical knowledge and his concentration on single aspects; his reading; his presentations to Portugal and Spain; and the significance for CC of d'Ailly's *Imago Mundi* and the marginal notes in CC's copy.

[438] **André Blum.** "Un but des voyages de Christophe Colomb," *Studi Colombiani* (Genoa: SAGA, 1952), 2: 211-16.

Reviews the documentary evidence suggesting that acquiring gold was the purpose of the first two voyages, eliciting support from the Genoese firm of Centurione. Points out that only in the reign of Charles V was this hope realized.

[439] **Ernesto Lunardi.** "L'Importanza del monastero di Santa María de la Rábida nella genesi della Scoperta d'America," *Studi Colombiani* (Genoa: SAGA, 1952), 2: 451-67.

Reviews the various important opportunities and influences provided to CC by the institution and the friars he met there.

[440] **Demetrio Ramos.** "Por que tuvo Colón que ofrecer su proyecto a España," *Rev Indias* (Madrid), 31, nos. 125-6 (1971): 77-137.

Rpt., Cuadernos Colombinos no. 3 (Valladolid: Casa-Museo de Colón, 1973). CC failed to get a commission to sail west to the Indies from John II of Portugal in 1484 because the only practicable route in CC's mind was in the latitude of the Canaries, a latitude that John wished to avoid at all costs to avoid breaking the Treaty of Alçacovas and thus inviting the Spanish to invade the space along the West African coast that enabled the Portuguese to plan and execute voyages south around Africa. When CC went to Spain he had not broken definitely with John II and might well have continued the negotiations on the basis of another route or turned to France or England; but when he reached La Rábida and learned, from Pedro Vásquez's account of bird-flights, that there must be lands to the southwest of the Azores, he decided to make his presentation to Spain, which controlled the necessary trade-wind route starting at the Canaries.

[441] **Paolo Emilio Taviani.** *Cristoforo Colombo: la genesi della grande scoperta.* Novara, Italy: Istituto Geografico de Agostini, 1974. 2 vols.

Biography of CC to the 1492 landfall, to complement Morison's *AOS* (#132), emphasizes CC's life ashore. Vol. 1 contains 44 chapters comprising (1) a popular narrative biography and (2) intercalated accounts of the history, geography, and biology of the successive locales visited by CC. For each of these 44 chapters, Vol. 2 provides (1) a corresponding chapter of learned discussion addressing selected issues and (2) extensive bibliographies. The indexes treat (a) names and authors and (2) places. These books are among the most handsome ever devoted to CC. See also #252.

[442] **Emiliano Jos.** *El Plan y la Génesis del descubrimiento colombino,* ed. Demetrio Ramos. Valladolid: Casa-Museo de Colón, 1980. 140 pp.

Cuadernos Colombinos no. 9. Essays based on a course of lectures summarizing the author's scholarly opinions on the genesis of CC's Enterprise of the Indies.

Twofold purpose: (1) to describe CC's preparation, theoretical and practical, for his Enterprise; (2) to refute various false or misleading assertions respecting the development and purpose of the enterprise. An extensively documented study.

Published posthumously with many added notes by the editor.

4. Consequences Involving Columbus Himself

[443] **Rudolfo Barón Castro.** "The Discovery of America and the Geographical and Historical Integration of the World," *Cahiers d'Histoire Mondiale* (Neuchatel), 6 (1961): 809-832.

Traces the gradually intensifying European activity in the Atlantic from the Phoenecian contact with the Canaries to CC's 1492 voyage of discovery, which was a spinoff of the European desire to re-establish relations with the Far East after the overland route began to close up in the late 13th century. CC's career treated pp. 812-22. Beginning with the Portuguese voyages around Africa to India and Magellan's voyage across the Pacific, the Iberian nations wove a progressively tighter net around the globe; and the concept of discovery gradually changed, first into a concept of communication and then one of scientific research. Although other Europeans joined in the activity very early, the main lines were established and the essentials achieved during the dynamic period of the Iberian voyages.

[444] **Carl O. Sauer.** *The Early Spanish Main.* Berkeley: Univ. of California Press, 1966.

A geographer's history of the early Spanish discoveries and explorations in the Caribbean area to 1519. First half deals with CC's explorations and especially with the results of his mismanagement of Hispaniola. Emphasizes the tragic effect on the native population of the Spaniards' misunderstanding of native agriculture with its resourceful and productive mound-farming, which was destroyed by the invaders.

C. E. Nowell deemed this "the best book existing on the early Caribbean," *Hisp Amer Hist Rev*, 48 (1968): 693.

[445] **Alfred W. Crosby.** *The Columbian Exchange: Biological and Cultural Consequences of 1492.* Westport, CT: Greenwood, 1972. 268 pp.

Six chapters on the interchange of Old- and New-World organisms and the resultant changes in global ecology. Special attention to the exchange of epidemic diseases, food crops, and animals. The result is a much more homogeneous collection of life-forms, a permanent alteration that has left us with an impoverished genetic pool. "We, all of the life on the planet, are the less for Columbus, and the impoverishment will increase." (p. 219). Excellent essay on syphilis, pp. 122-64.

This is a vastly influential book, especially in its introduction of the term "exchange" as a substitute for the inadequate term "discovery."

[446] **Earl J. Hamilton.** "What the New World Gave the Economy of the Old," *First Images of America* (Berkeley: Univ. Cal Press, 1976), 2: 853-84.

Reviews the influence of New World fauna (very limited, mainly through turkeys and fur); flora (extremely imporant, esp. maize, potatoes, tomatoes, tobacco, cassava); Old World flora transplanted to the New (sugar cane, coffee, bananas); precious metals, stimulus of prosperity in Europe as long as the money supply stayed ahead of rising prices and enabled capitalists to amass fortunes for investment; and especially the expansion of Western culture and its resulting world dominance. Appendix on syphilis (879-880) reviews the onset in Europe and concludes that the devastating European epidemic that began in the 1490's almost certainly came from infected persons in CC's returning ships in 1493.

[447] **David Henige.** "On the Contact Population of Hispaniola: History as Higher Mathematics," *Hisp Amer Hist Rev,* 58 (1978): 217-37.

Addresses the destruction of native American population following 1492. Seeks to demonstrate that recent attempts to estimate the population of Hispaniola (and other lands discovered in the 15th and 16th centuries) are futile. Rejects as indeterminate the estimate of Sherburne Cook and Woodrow Borah (*Essays in Population History* [Berkeley: Univ. Cal Press, 1971-74], 1:376-406) that 8 million natives lived on Hispaniola when CC arrived. Cf. Wm. M. Denevan, *Native Populations of the Americas in 1492* (Madison: Univ. Wis. Press, 1976) and Henry F. Dobyns, *Native American Historical Demography* (Bloomington: Indiana Univ. Press, 1976).

[448] **Louis-André Vigneras.** "Diego Méndez, Secretary of Christopher Columbus and *Alguacil Mayor* of Santo Domingo: A Biographical Sketch," *Hisp Amer Hist Rev* 58 (1978): 676-96.

Treats three crucial aspects of the early colonial experience: (1) the connection between a spectacular achievement (Méndez's rescue of CC on Jamaica) and the life of the achiever; (2) The colonial and imperial measures that lay behind Méndez's rewards in property, honors, and office, as a paradigm of the development of the *encomienda*; and (3) the vagaries of Mendez's fortunes as a paradigm of the fortunes and misfortunes of ambitious Spaniards in the Indies at this time.

[449] **Marisa Vannini de Gerulewicz.** "L'America agli occhi dei primi scopritori," *Atti III Conv Internaz Stud Col* (Genoa: CIC, 1979), pp. 405-426.

A reconstruction of the initial impressions of the Europeans, with quotes from the reports of CC, Ferdinand C, and many other, somewhat later explorers. Emphasizes the difficulty caused by the lack of any adequate previous experience or conceptions, a lack that caused the new experiences to be assimilated quite erroneously to what was already known. This natural bankruptcy of adequate preparation delayed for a long time the adequate perception of what had been found.

C. CC'S LITERACY, LANGUAGE, AND WRITING

[450] **J. de Dios de la Rado Delgado.** "Tres autógrafos de Colón," *El Centenario* (Madrid), 3 (1893): 219-29.

Two studies: (1) a study of folio 59 verso of CC's *Book of Prophecies,* which records the famous quote from Seneca's *Medea,* "Venient annis . . . Ultima Tille." There follow CC's Castilian translation and astronomical observations *in re* Saona Island and Jamaica. Rado observes that CC's handwriting as seen here and elsewhere takes 3 forms: (a) a neat, careful style used for copying passages from other authors; (b) a less formal but still neat style for translating and for setting down fully-formed thoughts; (c) a completely cursive and much less legible style for recording the swift flow of thoughts not yet put in order. R. concludes that the whole page is written by CC. The quote from Seneca is in style (a); the translation is (a) shading to (b); and the astronomical comments that follow, in style (c). Rejects opinion of Simón de la Rosa that the quote from Seneca is in another hand.

Study (2) examines two letters from CC to Fr. Gorricio, the originals of which were discovered by Angel González in Guatemala in 1892. These are the letters, from Seville, of 21 Mar. 1502 and 27 Dec 1504. They had already been published by Navarrete in 1825. Establishes that the 2 letters had been in a codex from which Spotorno had copied his *Codice Diplomatico*, #1 (1823) and Navarrete his *Colección de viajes*, #2 (1825). Some time later, they were removed and sent to Guatemala.

[451] **William Eleroy Curtis.** "The Existing Autographs of Christopher Columbus," *Annual Report of the American Historical Association, 1894* (Wash DC: AHA, 1895), pp. 445-518.

Records 29 complete letters and manuscripts entirely written in CC's hand. Translations of these items and 5 other by José Ignacio Rodríguez.

Mug 19

[452] **John Boyd Thacher.** "The Handwriting of Columbus," *Christopher Columbus* (New York: Putnam, 1904), 3: 84-488.

Reproduces (photographically) 42 letters and documents bearing the real or presumed handwriting of CC. In #81.

[453] **Robert Park.** "Columbus as a Writer," *Hisp Amer Hist Rev*, 8 (1928): 424-30.

Cites and quotes extracts in English translation from Hakluyt Society publications. CC's writings are highly interesting, full of vivid description, zeal for the faith, and optimism, but sometimes fanciful, as with his suggestion that South America was an apocalyptic discovery, and the Orinoco the Edenic fourth river. CC is a learned man, thoroughly acquainted with scripture and a bevy of classical and medieval authors.

Mug 83

[454] **F. Streicher.** "Die Kolumbus Originale: eine paleographische Studie," *Spanische Forschungen der Görresgesellschaft* (Munich), 1 (1928): 198-249.

A scholarly study of the marginal notes by CC and Bartholomew C in the books at the Biblioteca Colombina in Seville.

C. E. Nowell

[455] **Cecil Jane.** "The Question of the Literacy of Columbus in 1492," *Hisp Amer Hist Rev*, 10 (1930): 500-516.

Argues that since there is no CC autograph earlier than 1497 except some doubtful postils, CC was probably unable to write in 1492 and remained so until Bartholomew came to Hispaniola and taught him to write Spanish.

[456] **Ramón Menéndez Pidal.** "La lengua de Cristóbal Colón," *Bulletin Hispanique*, (Paris), 42 (1940): 5-27.

Rpt. in *La lengua de Cristóbal Colón, el estilo de S. Teresa, y otros estudios* (Madrid: Espasa-Calpa, 1942). Bases his analysis on the now largely rejected hypothesis that CC wrote his first postil in 1481 (i.e., De Lollis's #858 in the 1892 *Raccolta* ed. of Pius II's *Historia Rerum* [Venice, 1477]).

Proposes (1) that CC did not learn to write Italian in his boyhood, but only (possibly) some commercial Latin, and only spoke Genoese; (2) that in Portugal, 1476-85, CC learned to speak Portuguese and to write Portuguese Castilian full of "lusismos"; (3) CC continued to speak with this "lusismo" throughout his life under the Spanish flag; (4) his writings during the Spanish period are fluent, his vocabulary enormous and resourceful, though always heavily influenced by Portuguese. Cf. #s 462 and 468.

[457] **Julio F. Guillen Tato.** *La parla marinera en el Diario del primer viaje de Cristóbal Colón*. Madrid: Istituto Historico de Marina, 1951. 142 pp.

Notes that CC's locutions and vocabulary in the *Journal* strongly reflect lengthy sea experience. The frequent Navarrese locutions emphasize the strong influence of Cantabria and Galicia on Castilian mariners' language in the 15th century. Abstracted in #12, 2: 291-93.

[458] **Jaime Colomer Montset and Pedro Catalá Roca.** "Los escrituras de Cristóbal Colón y consideraciones sobre sus firmas," *Studi Colombiani*, (Genoa: SAGA, 1952), 2: 181-205.

Identifies 92 documents bearing writing by CC, and classifies the signatures therein into 7 groups to expedite further study.

[459] **Pedro Catalá Roca.** "Sobre los italianismos en la carta de Colón a Santangel," *Studi Colombiani*, (Genoa: SAGA, 1952), 2: 283-90.

Rejects D.R. Cuneo-Vidal's assertion (Cristóbal Colón: Genovese. Barcelona: Maucci, 1929) that frequent Italianisms occur in the *Letter* by refuting Cuneo's arguments for 18 locutions supposedly influenced by Italian.

[460] **Albert Tonneau.** "L'Énigme des chiffres de Christophe Colomb,: *Studi Colombiani* (Genoa: SAGA, 1952), 2: 137-80.

Describes some 18 variants of CC's curious signature and summarizes the interpretations of Spotorno, Charton, Sanguineti, Roselly de Lorgues, Dogné, Luis Ulloa, de Hevesy, Ribiero, Cardoso, Pestana, Amzalak, Wasserman, Streicher, Santos Ferreira and Ferreira de Serpa, and adds his own reading.

[461] **V. I. Milani.** *The Written Language of Christopher Columbus* (Buffalo NY: Forum Italicum, 1973). 152 pp.

Supplementary publication of *Forum Italicum* (SUNY – Buffalo), June, 1973. An alphabetically arranged record and discussion of words, used by CC in his Spanish writings, which anticipate the first recorded use of these words in dictionaries of the Spanish language. Cf. #462, #464, and #468.

[462] **Joaquín Arce Fernández.** "Problemi linguistici inerenti nel Diario de Cristoforo Colombo," *Atti I Conv Internaz Stud Col* (Genoa: CIC, 1974), pp. 52-75.

Analyzes the language of CC's *Journal of the First Voyage*, noting the judicious care of the abstractor Las Casas and the influence of CC's native Genoese and of his contact with Portuguese on the navigator's locutions. Rejects much of the analysis of Menendez Pidal (#456), and virtually all of that of Milani (#461). Cf. #468.

[463] **Giuseppe Caraci and I. Luzzana Caraci.** "Il Latino di Colombo," *Atti I Convegno Internazionale di Studi Americanistici* (Genoa: Assoc. Ital. Americanistici, 1976), pp. 87-93.

CC's Latin is mostly conditioned by his original language, Genoese, though also by the Lingua Franca of the seamen, as well as by his later use of the Spanish and Portuguese languages.

[464] **O. Chiareno.** "Recenti studi sulla lingua scritta di Colombo," *Atti I Convegno Internazionale de Studi Americanistici* (Genoa: AIA, 1976), pp. 107-117.

Rpt. in *La lingua di Colombo e altri scritti di americanistica,* (Genoa: Di Stefano, 1988), pp. 7-25. A survey, with special attention to Virgil Milani, *The Written Language of CC*, #461, and Joaquín Arce's edition, with Gil Esteve, of CC's *Diario*, #51.

[465] **Yakov Malkiel.** "Changes in the European Languages under a New Set of Sociolinguistic Circumstances," *First Images of America* (Berkeley: Univ. Cal Press, 1976), 2: 581-93.

An exercise in isolating "the universals that can be abstracted from the separate records of English, Dutch, Scandinavian, French, Spanish, Portuguese, and Italian speech projected onto the soil of the Western hemisphere" (p. 590). In tracing the amalgamated language of Spanish emigrants who mixed with Portuguese and Italians, sees the language of CC as "nothing short of paradigmatic. This native of Genoa . . . used a Castilian generously interspersed with echoes of his earlier experiences and exposures" (p. 587).

[466] **Paola Navone.** "Colombo e il 'Bestiario' dell'oriente meraviglioso," *Columbeis I* (Genoa: IFCM, 1986), PP. 117-23.

CC's anomalous spellings in recording the names of animals in his postils seem to derive not so much from his Genoese background as from spellings used in his bestiary.

[467] **Stefano Pittalugo.** "Il 'vocabulario' usato da Cristoforo Colombo (Una postilla all' 'Historia rerum' di Pio II e la lessicografia medievale)." *Columbeis I* (Genoa: IFCM, 1986), pp. 107-115.

CC's postils reflect his use of the "Vocabulario" of Giovanni Balbi and Balbi's habit of comparing Latin texts with recent translation of the texts; but the postils do not reflect medieval exegetical habits so much as they reflect a lively curiosity that makes CC's notes more like the remarks with which the humanists indicated notable passages in the margins of their texts.

[468] **Giorgio Bertoni.** "Appunti sugli italianismi linguistici di Colombo," *Columbeis II* (Genoa: DARFICLET, 1987), pp. 19-29.

Reviews in detail some of the literature on CC's language, e.g., R. Menéndez Pidal, *La Lengua de CC* #456; J. Arce Fernández, "Problemi linguistici inerenti in il *Diario* di CC," # 462; Arce's parallel comments in the edition of the *Diario* #51; and V. Milani, *The Written Language of CC*, #461.

Suggests a nautical and Mediterranean origin for some of the terms rather than a landbound one. This reflects Bertoni's conviction that C. Varela (*Cristóbal Colón: Textos y documentos completos*, #54) is correct in appraising CC's gift of language as one that enables him to choose words easily understood, drawing upon the various and hybrid sources available from his nautical experience.

[469] **Giorgio Bertoni.** "L'occhio, l'ancora, la scrittura, lo sguardo dell'almirante," *Columbeis II* (Genoa: DARFICLET, 1987), pp. 153-80.

It is not likely, as Todorov maintains (#471), that CC did not have the notion of linguistic plurality (and the idea of the "other"). It is more likely that the *Journal* was not addressed to posterity, to us, but to the Spanish monarchs and with the purpose of presenting himself as the sole protagonist of the Enterprise of the Indies: the one who accomplished the feat and the one who had the vision in the first place. B. analyzes CC's use of the pronoun "I" and the verb "see," and the procession of signs that are always confirming his vision and his promises to the Spanish monarchs.

[470] **Mario Damonte.** "Le lingue di Cristoforo Colombo," *Columbeis II* (Genoa: DARFICLET, 1987), pp. 9-18.

Even though CC spoke Genoese and Marinaresco and probably knew Tuscan and Portuguese, still the only language he wrote, except for a few notes in Latin and Tuscan, was Castilian. D. addresses the Castilian and the words that have Portuguese and/or Genoese origin, and concludes that though much has been done, the last word has not been said on CC's language.

[471] **Anna M. Mignone.** "Index verborum Columbianus: Il *Diario di Bordo*," *Columbeis II* (Genoa: DARFICLET, 1987), pp. 41-102.

Presents a concordance of the *Journal*, the first completed section of a lexical concordance of CC's works directed by Francesco Della Corte of the Univ. of Genoa. The index is based on the De Lollis text in the 1892 *Raccolta* (#46). An essay detailing the criteria for computer analysis, pp. 47-50, precedes the concordance.

[472] **Stefano Pittalugo.** "Cristoforo Colombo amanuense (e il suo incunabolo del 'Catholicon' di Giovanni Balbi)," *Columbeis II* (Genoa: DARFICLET, 1987), pp. 137-51.

The notes CC wrote in his books, whether summaries, comments, or references to other sources, were always intended for personal use. His Latin, though as problematic as the man who used it, was neither Genoese nor Iberian; neither a secret language nor the confused babbling of a child. If a secret exists in his language, it is the private and "so humanistic" way he reads, always looking for prophecies or confirmation of his project.

[473] **Rosanna Rocca.** "Il lessico di Michele da Cuneo," *Columbeis II* (Genoa: DARFICLET, 1987), pp. 225-30.

An analysis of the diction of Michele da Cuneo's letter, intended to give an understanding of how two Ligurians of the time, both from Savona, and both of noble family, communicated with each other.

D. CHARACTER, THOUGHT, AND PERSONALITY OF COLUMBUS

[474] **Giuseppe Portigliotti.** "Per una biografia psicologica di Cristoforo Colombo: I. Le idee messianiche. II. I vaticini profetici," *Annali dell'Ospedale Psichiatrico della Provicia di Genova* (Cogoleto), 3 (1932), 1-32.

Using CC's own writings, in particular the *Libro de las profecías*, shows how CC felt especially cared for by Providence and especially chosen for 3 missions: (1) to spread the spirit of God to the extreme limits of the earth; (2) to deliver the Holy Land from servitude to Mohammed; (3) to prepare for the descent of the Holy Spirit on an Earth united and redeemed by the light of Christ.

[475] **E. G. R. Taylor.** "Idée Fixe—the Mind of Columbus," *Hisp Amer Hist Rev*, 11 (1931), 289-301.

CC, neither a rational thinker nor a professional sailor, was a man governed by the idea of his destiny to discover and rule rich unknown lands, which (since they were not in Europe or Africa) he inevitably placed in Asia, where a Genoese would assume all wealthy lands to be.

[476] **Armando Alvarez Pedroso.** *Cristóbal Colón: Biografía del Descubridor.* Havana: Cultural, 1944. 498 pp.

Emphasizes especially the author's conception of CC's personality. Sections: (1) Hombre (2) Genio (3) Místico. A fourth section, "Parte analytico," comprises essays on a series of CC topics.

The book is synthesized from updated versions of a number of separate publications, not all listed separately in this *Guide* because they are treated in final form in this book. These include "CC: Hombre-genio-místico"; "Plan científico de Colón para el descubrimiento de América"; "CC no fue Hebreo" (cf. #651); "Los restos mortales del descubridor" (cf. #691); and "El verdadero retrato de CC."(cf. #684).

[477] **Ramón Iglesia.** "El hombre Colón," *El hombre Colón y otros ensayos* (Mexico City: El Colegio de Mexico, 1944), pp. 17-49.

A reading of CC's personality as that of a hard, practical Genoese whose supposed mysticism in his final years was simply another means of pursuing the honors, influence, and property that he had been promised in 1492.

[478] **Torquato C. Giannini.** "Psicologia Colombiana: base psicologica del disegno di Colombo," *Studi Colombiani* (Genoa: SAGA, 1952), 2: 113-118.

We must recognize that CC was not merely obsessed with the conviction that he would succeed in his voyage, but that his mind was prepared for the voyage by developing the Genoese and Italian tendency to look West across the ocean for the means to redress the loss of trading advantages in the Eastern Mediterranean and the Black Sea. CC's preparation came from the tradition that produced the Vivaldi voyage, the Toscanelli map, and many other consonant Italian developments.

[479] **Charles Verlinden.** "Christophe Colomb: esquisse d'une analyse mentale," *Rev Hist América* (Mexico City), no. 89 (1980), pp. 9-27.

Reviews CC's whole career (without deep concern for details; he thinks CC's presence on board a Genoese fleet in April [sic] 1476 is certain) and then notes CC's vigorous adaptability and great capacity for developing his talents, as in becoming a great navigator and capable commander not only of a ship but a fleet. As further important traits, points out his dignity and his perseverance in adversity, the last of which was the gift that made his achievement possible This was his genius: not his adaptability and power to learn but the gift of will with which he stuck to an idea that though wrong was very fecund.

[480] **Paolo Emilio Taviani.** "Columbus the Man: A Psychologically Modern Man of the Middle Ages," *Columbus and his World* (Ft. Lauderdale FL: CCFL, 1987), pp. 1-12.

Reviews the concrete practicality of CC's nature, which anticipated the modern Western habit of mind that has gradually embraced the whole world.

[481] **Juan Gil.** *Mitos y utopías del descubrimiento: 1. Colón y su tiempo.* Madrid: Alianza, 1988. 302 pp.

This study concentrates in great part on the development of CC's ideas about his enterprise and its results, and his responses to the criticism heaped upon him by his contemporaries. The book constitutes a major attempt to delineate the psychology of the discoverer, and develops Gil's frequently asserted thesis about the dating of most of CC's postils in the 1490s. Chapters on CC: 1. Los enseños del primer viaje: el oriente según Colón; 2. La euforia índica y las primeras decepciones; 3. Crisis del prestigio colombino; 4. Colón a la defensiva (develops Gil's ideas about CC's postils, cf. #s 77 and 78); 5. Mas censuras eruditas; deposición del Almirante; 6. Nueva crítica y último viaje: la aurea salmonica; 7. La religiosidad de CC.

These 7 chapters are followed by two on subsequent developments: 8. Los eruditos frente al dilema ofírico: la *translatio imperii*; 9. El fin de un mito: la fuente de la juventud.

[482] No entry 482.

E. COLUMBUS AND NATURAL SCIENCE

1. General Studies

[483] **J. Rey Pastor.** *La ciencia y la técnica en el descubrimiento de América.* Buenos Aires: Espasa-Calpe Argentina, 1945. 171 pp.

A survey of the subject. Chapters: 1. Introduction. 2. Cosmography, cosmology, and mathematics. 3. Botany, mining, metallurgy. 4. Miscellaneous.

[484] **M. Tenani.** "Affirmazioni de Colombo circa gli uragani definitivamente confermate dopo 462 anni," *Bol Civico Ist Col* (Genoa), 2, no. 4 (1954): pp. 36-40.

Hurricane "Alice," in Jan 1955, confirms CC's description of a hurricane in Feb 1493, following the same trajectory. The hurricane of Sept 1928 confirms CC's observation of the hurricane of July 1502.

[485] **German O. Galfrascoli.** "Náutica y ciencias geográficas en la época de Colón," *Atti III Conv Internaz Stud Col* (Genoa: CIC, 1979), pp. 235-56.

A survey, first of nautical science in CC's day, then of geographical science.

[486] **John Winter.** "San Salvador in 1492: Its Geography and Ecology," *Columbus and his World* (Ft. Lauderdale FL: CCFL, 1987), pp. 313-320.

CC in 1492 could have experienced the climax evergreen woodlands of San Salvador, when he writes of very green trees, fruits, and much water. Unfortunately, all the woodland has now been cleared, burned, and selectively cut at least once. In addition, the coastal topography has undergone several changes through progradation, retrogradation, and sedimentation. This has created an island which is a far different sight from that of CC's Guanahani. A reconstruction of the island as it was is made possible through historical records and environmental processes.

2. Cosmology and Cosmography

a. General Studies

[487] **Vittore Bellio.** "Notizia delle più antiche carte geografiche che si trovano in Italia riguardanti l'America," *Raccolta,* IV.2 (1892), 101-221.

An account of 31 maps, in 12 chapters. Tables, maps.

[488] **Henry Harrisse.** "Cartographia Americana Vetustissima," in *The Discovery of North America, A Critical, Documentary, and Historic Investigation* (London: H. Stevens, 1892; rpt. Amsterdam: N. Israel, 1961), pp. 363-648.

A chronological account and description of 239 cartographical documents, 1461-1536. Executed with Harrisse's usual thoroughness, paralleling the *BAV*, #731.

[489] **Juan Valera.** "Concepción progresivo del nuovo mondo," *El Centenario* (Madrid), 3 (1893): 145-55.

Sets forth with clarity and authority the development of the concept of the New World. Cf. John Fiske, *The Discovery of America*, #128.

[490] **John Boyd Thacher.** *The Continent of America: Its Discovery and Baptism.* New York: Benjamin, 1896. 270 pp.

"An essay on the nomenclature of the old continents; a critical and bibliographical inquiry into the naming of America and into the growth of the cosmography of the new world; together with an attempt to establish the landfall of Columbus on Watling island, and the subsequent discoveries and explorations on the mainland by Americus Vespucius." [the continuation of the title]

[491] **Louis Salembier.** "Pierre d'Ailly and the Discovery of America," *Historical Records and Studies of the U. S. Catholic Society*, 7 (June 1914): 90-131.

Sets forth d'Ailly's geographical ideas about sailing west to India, d'Ailly's sources for these, and their influence on CC. Supports Vignaud's contention that CC did not know d'Ailly until Bartholomew C. brought the book from France after the 1st voyage.

Mug 87

[492] **William H. Babcock.** "Antilia and the Antilles," *Geog Rev* (NY), 9 (1920): 109-124.

Proposes that the legendary name "Antilia" as represented on Beccario's 1435 map is really the "Queen of the Antilles," i.e., Cuba. Reply by G. R. Crone, #505.

C. E. Nowell

[493] **George E. Nunn.** "The Lost Globe Gores of Johann Schöner," *Geog Jour* (London), 60 (1921): 476-80.

The 1523 globe gores attributed to Schöner (now lost) show an attempt to reconcile Magellan's discoveries with CC's ideas about South America as an appendage of Asia.

[494] **Edward Heawood.** "A Hitherto Unknown World Map of A.D. 1506," *Geog Jour* (London), 62 (1923): 279-93.

Makes a detailed analysis of the Contarini map purchased by the British Museum in 1923, dated 1506, and hypothesizes that this may be the earliest world map to reflect CC's discovery of the West Indies (Cuba, Hispaniola, the Lesser Antilles as a group, and the Venezuelan coast are clearly represented). Map reproduced following p. 320.

[495] **George E. Nunn.** *The Geographical Conceptions of Columbus.* New York: American Geographical Society, 1924. 148 pp.

Includes 4 studies: (1) CC's determination of the length of a terrestrial degree: especially valuable for demonstrating that CC measured the degree on a *meridian*; (2) Evidence on the first voyage of CC's knowledge of winds & currents; (3) whether CC believed he had reached Asia on the 4th voyage (concludes he did); (4) the identity of "Florida" on the Cantino map of 1502 (i.e., not Florida).

[496] **Roberto Almagià.** "Le prime conoscenze dell'America e la cartografia italiana," *Atti XXII Congresso Internazional degli Americanisti, 1926* (Rome, 1928), 2: 589-92.

Calls for the establishment and publication, in Italy if possible, a raccolta of maps and other important cartographic documents relating to the knowledge of America since the beginning of the 17th c, within such an historical context as to make it possible to delineate the cartography of the Americas prior to 1600.

[497] **Cecil Jane.** "The Opinion of Columbus concerning Cuba and the 'Indies,'" *Geog Jour* (London), 73 (1929): 266-70.

Suggests that CC's action in forcing his crew to swear that they thought Cuba to be the mainland (during the exploration of the south coast of Cuba, 1494) was a cautionary device to keep the crew from telling the other Spaniards in Hispaniola that they had only discovered more islands, and thus to further discredit him with his subjects in Hispaniola.

[498] **George E. Nunn.** *The Columbus and Magellan Concepts of South American Geography.* Glenside CA: Nunn, 1932.

Analysis of maps drawn between 1420 and 1425 shows how mapmakers accommodated CC's discoveries to the land masses as conceived before 1492. Magellan's voyage did not prove to these mapmakers that South America was not a part of Asia, and Magellan seems to have thought it was.

Mug 75

[499] **G. H. Kimble.** "Portuguese Policy and its Influence on Fifteenth Century Cartography," *Geog Rev* (NY), 23 (1933): 653-59.

The Portuguese royal policy of concealing maps and records of voyages of discovery impeded the transcription of their discoveries onto maps and thus impeded the discipline of cartography.

[500] **Armando Cortesão.** *Cartografía e cartógrafos portugesos dos secolos xv e xvi.* Lisbon: Seraa Nova, 1935. 2 vols.

Chapter 4, "Bartolomeu e Cristovão Colombo cartógrafos." Sections on the "La Roncière" map at Paris, pp. 238-42, and the Piri Re'is map, pp. 242-48.

[501] **Richard Uhden.** "Die antiken Grundlagen der mittelalterlichen Seekarten," *Imago Mundi*, (Leiden) 1 (1935-36): 1-20.

The sea charts of the Middle Ages show their ancient derivation in the wind rose, employed by the ancient Greek Timosthenes, and in a world map in Ravenna as early as the 7th century.

Further developed by E. G. R. Taylor, *Imago Mundi* (Leiden), 2 (1936-37): 23-26.

[502] **George E. Nunn.** "Martinus of Tyre's Place in the Columbus Concept," *Imago Mundi* (Leiden), 2 (1936-37): 27-35.

Martinus' only influence on CC's geography was the placing of Cattigara at longitude 225 deg. E.

[503] **Paolo Revelli.** *Cristoforo Colombo e la scuola cartografica genovese.* Genoa: Stabilimenti Italiani Arti Graphichi, 1937. 563 pp.

A monumental study, profusely and munificently illustrated and documented, of the relation between CC's Enterprise of the Indies and the tradition of mapmaking evolved in Genoa beginning in the 12th century. Infers that CC's idea for his western voyage grew from the ambience of this cartographical tradition. Includes extensive illustrated study of CC's Genoese background and culture.

[504] **Ernst Zinner.** *Kolumbus und die "Ephemerides" des Regiomontanus.* Gotha: Petermanns, 1937. 367 pp.

Although this book has not been located at publication time, there is much reason to believe that it does exist.

[505] **G. R. Crone.** "The Origin of the Name 'Antillia,'" *Geog Jour* (London), 91 (1938): 260-61.

Answers Wm. H. Babcock, #492. What Babcock thought an authentic island was probably a displaced representation of the Gates of Hercules. The legend "newly reported islands" on Beccario's 1435 map refers to the Azores.

C. E. Nowell

[506] **Demetrio Rámos Pérez.** "Problemas geográficos de las navegaciones colombinos," *Boletín de la Real Sociedad de la Geografía* (Madrid), 84 (1948): 368-96.

Examines a series of problems connected with (1) the circulation of the Atlantic currents, (2) the Sargasso Sea, (3) compass declination, and (4) the shape of the bordering land and the contrasting climates, all of which had tended in medieval times to isolate America even though the technical sailing problems had been solved. Concludes that CC, besides being an audacious navigator, was an imaginative geographer who came to grips successfully with these problems as well as with many of the related human problems.

[507] **Milton V. Anastos.** "Pletho, Strabo, and Columbus," *Annuaire de l'institut de philologie et d'histoire orientales et slaves* (Brussels), 12 (1952): 1-18.

Traces the introduction of Strabo's ideas into the West. Rejects the notion that Strabo's conception of "oikoumene" might have reached Western Europe before 1400. Pierre D'Ailly sometimes echoes Strabo, as in a suggestion that in a fair wind a ship could cross the ocean from Spain to India.

Martin Torodash

[508] **Arthur Davies.** "Origins of Columbian Cosmography," *Studi Colombiani,* (Genoa: SAGA, 1952), 2: 59-67.

Nowell (#435) notwithstanding, CC's cosmographical conceptions were Italian, not Portuguese. CC, adopting Toscanelli's estimate that China was 1650 leagues west of Europe, was rejected by John of Portugal (1484), by the Spanish crown (1486) and by Henry VII of England (1488), because the voyage was too long. Finally, reading D'Ailly's *Imago Mundi* with his brother Bartholomew in 1489, he found an argument for a degree of 56 2/3 miles. Without self-deception he decided to use a degree of 56 2/3 Italian miles simply in order to reduce his estimate of the distance to 750 leagues. This made the voyage affordable and gained approval from the Spanish crown. But he knew all the time he could not reach Asia this way.

[509] **Johannes Keuning.** "The History of Geographical Map Projections until 1600," *Imago Mundi* (Leiden), 12 (1955): 1-24.

Reviews not only the commonly used equidistant and stereographic projections but also a series of projections that constitute a brief history of the subject from its beginning: azimuthal or zenithal projections, both equidistant orthographic, gnomonic or gnomic, and stereographic; Ptolemaic projections, 1st and 2nd; cordiform; that of Martinus of Tyre; Mercator; trapezoidal; globular; oval; and the separate projections of L. da Vinci, Orontius Finaeus; Guillaume le Teste; and Jean Cossin.

[510] **A. Gamir Sandoval.** "Posible evolución en el pensamiento geográfico colombino (1492-1506)," *Rev Indias* (Madrid), 20, nos. 81-82 (1960): 31-63.

Contrary to the position taken by Vignaud, CC set out in 1492 expecting to reach Asia; but gradually, over the years, must have become conscious that he had discovered new lands distinct from Asia—though still, in his view, close to Asia.

Includes a compendious documented account of the experience of the other mariners between 1492 and 1506.

[511] **S. García Franco.** "Sobre un portulano de 1500. Lo dibujó Colón?" *Rev Gen Marina* (Madrid), 158 (1960): 27-29.

Takes issue with A. Wilhelm Lang, whose article in *Imago Mundi* (Leiden), 12 (1955) reproduces on p. 32-33 part of a "Portuguese world map attributed to CC, ca. 1500"; the original is in the Bibliotheque Nationale, Paris. García takes issue with Lang's statement that the map is not by CC; G. points out that on this map the degree at the equator is 14 1/6 leagues, corresponding to the 56 2/3 miles that CC used, adopting it from Alfraganus. G. tests various latitudes on the map in question; and, finding the degree virtually the same, offers this as strong evidence that the map is by CC.

[512] **Carlos Sanz.** *El descubrimiento de América. Los tres mapas que lo determinaron.* Madrid, 1972.

Translated as an article in *Terrae Incognitae* 6 (1974): 77-84. Studies the map attributed to Claudio Ptolomeo, that by Henricus Martellus ca. 1490, and the 1507 planisphere that named America for the first time.

María J. Sarabia Viejo

[513] **Osvaldo Baldacci.** "La cartonautica medioevale precolombiana," *Atti I Conv Internaz Stud Col* (Genoa: CIC, 1974), pp. 121-36.

Part 3, pp. 132-36, speculates on the inferences to be drawn from CC's postil # 859 (De Lollis no.) to Piccolomini's *Historia Rerum,* which comprises a geographic projection of a planisphere.

[514] **Hildegard Binder Johnson,** "New Geographical Horizons: Concepts," *First Images of America* (Berkeley: Univ. of California Press, 1976), 2: 615-33.

Explains the usefulness of maps and other book illustrations as a means of tracing the development of fruitful geographical concepts such as the irregular distribution of land on the globe, the world-ocean as a link rather than a barrier, the global region or zone, general vs. special geography, and the watershed.

[515] **Ursula Lamb.** "Cosmographers of Seville: Nautical Science and Social Experience," *First Images of America* (Berkeley: Univ. of California Press, 1976), 2: 657-86.

The *Casa de Contratación,* first public lay institution in Europe for science and technology, was formed in response to demands for improving the nautical link with the New World. Its efforts were supplemented by the Consejo Real y Supremo de las Indias and the Royal Academy of Mathematics. Their choices of problems to be studied were strongly influenced by social experience, and shed light on the role of social restraint in shaping the physical sciences and in evolving the scientific professions.

[516] **Francis M. Rogers.** "Celestial Navigation: From Local Systems to a Global Conception," *First Images of America* (Berkeley: Univ. of California Press, 1976), 2: 687-704.

Beginning at the dawn of the Age of Discoveries, navigators gradually moved from a mere knowledge of local systems to a global conception of celestial navigation. Through the cumulative employment of the sextant, the chronometer, the nautical almanac, charts and chart tools, and sight-reduction tables, man has been freed to roam the seas and skies independent of shore installations.

[517] **Norman J. W. Thrower.** "New Geographical Horizons: Maps," *First Images of America* (Berkeley: Univ. of California Press, 1976), 2: 659-74.

"Renaissance cartography reflects remarkable originality in symbolization, projection, and reproduction techniques while, at the same time, delineating those fundamental changes in knowledge of the earth which resulted, particularly, from the European discovery of the New World" (p. 659).

[518] **Caterina Barlettaro and Ofelia Garbarino.** "Il fondo cartografico conservato nell'archivio di stato di Genova," *Atti II Conv Internaz Stud Col* (Genoa: CIC, 1977), pp. 181-88.

Broadly delineates the collection of some 200 documents, mostly maps, explored by Emilio Marengo and currently being put in order by Aldo Agosto, which will be valuable in reconstructing the geographic and cartographic culture from which CC came.

[519] **Elio Migliorini.** "Gli studi colombiani nell'opera dei geografici italiani dell'ultimo secolo," *Atti II Conv Internaz Stud Col* (Genoa: CIC, 1977), pp. 17-71.

Reviews the achievement of the geographers of the 1892-96 *Raccolta* (Hugues, Amat di S. Fillipo, Pennesi, Bellio, Bertelli) and of their 20th-century successors, notably Alberto Magnaghi, Paolo Revelli, Roberto Almagiá, and Giuseppe Caraci. Notes a steadily evolving sense of the necessary scientific objectivity in these studies.

[520] **I. Luzzana Caraci.** "Colombo e le longitudini," *Boll Soc Geog Ital* (Rome), Ser. 10, 19 (1980): 517-29.

CC used longitude not so much for navigation as to fix the position of the newly discovered lands in the general view of his geographical conceptions. He made at least two attempts to measure longitude, one at Saona Island in 1494 and the other in Jamaica in 1504.

[521] **Juan Gil.** "Pedro Mártir de Anglería, intérprete de la cosmografía colombina," *Anuario Estud Amer* (Seville), 39 (1982): 487-502.

Addresses the passage in the *Decades* treating the exploration of the south coast of Cuba, the beginning of which CC named "Alpha and Omega." Concludes that CC's cosmography was in the service of a religious cause. CC went searching for the new heaven and the new earth, and this is what he found.

[522] **Osvaldo Baldacci.** "La geocarta come documento storico colombiano," *Pres Ital Andalu II* (Bologna: Cappelli, 1986), pp. 157-68.

An account of CC's technique in projecting his planisphere, and a speculation on its usefulness as an instrument of navigation and discovery.

[523] **Graziella Galliano.** "Forma e dimensioni della terra nelle postille colombiani," *Miscellania I* (Genoa: University of Genoa, 1986), pp. 175-91.

A content analysis of CC's marginal notes in D'Ailly's *Imago Mundi* and Pius II's *History* shows three related cosmological subjects: the form of the land and of the whole earth; the size of the land and the earth; and the division into climatic zones. Hypothesizes that, because of an original note on the *cuadernillo* in Pius II's *History* whose *terminus quo* is 1485, CC's notes in both volumes are *prior* to 1486.

[524] **Rosanna Rocca.** "Colombo e la 'Isla de Córcega,'" *Columbeis I* (Genoa: IFCM, 1986), pp. 133-37.

CC's labeling of the easternmost point in Cuba as *aurea* might not be a reference to the Golden Chersonese of the Indian Ocean, with which CC connected Cuba; it might just reflect a resemblance to a promontory that CC probably knew on the island of Corsica.

[525] **Georges A. Charlier.** "Value of the Mile Used at Sea by Cristóbal Colón During his First Voyage," *Columbus and his World*, (Ft. Lauderdale FL: CCFL, 1987), pp. 115-120.

Columbus used the Roman mile of about 1480 meters.

[526] **Tony Campbell.** "Portolan Charts from the Late Thirteenth Century to 1500," *Cartography in Prehistoric, Ancient, and Medieval Europe and the Mediterranean*, ed. J. B. Harley and David Woodward (Chicago: U. Chicago Press, 1987), pp. 371-463.

Exhaustive, thoroughly illustrated and documented study of the subject.

[527] **David Woodward.** "Medieval Mappaemundi," *Cartography in Prehistoric, Ancient, and Medieval Europe and the Mediterranean*, ed. J. B. Harley and David Woodward (Chicago: U. Chicago Press, 1987), pp. 286-368.

Exhaustive, thoroughly illustrated and documented survey of the subject.

b. Columbus and Toscanelli

[528] **G. Uzielli.** *La vita e i tempi di Paolo dal Pozzo Toscanelli. Raccolta*, V.1 (Rome: Ministero della Pubblica Istruzione, 1894). 745 pp.

Biography set within elaborate historical background, concluding with inference that T. opened the way for CC, Vespucci, Cabot,

Verrazzano, and all the European navigators who explored the world in the succeeding years.

Chap. 6, pp. 308-85, by Giovanni Celonia, discusses T's astronomical work.

Plate 10, Uzielli's reconstruction of the map of Toscanelli, based on the data in the *Codici Magliabechiano* (MS in the Florentine Library, class xi, no. 121), with reference to Fra Mauro's planisphere and Behaim's globe.

[529] **Henry Vignaud.** *Toscanelli and Columbus: The Letter and the Chart of Toscanelli on the route to the Indies by way of the west, sent in 1474 to the Portuguese Fernam Martins, and later on to Christopher Columbus,* etc. London: Sands, 1902. 365 pp.

Title continued: A critical study on the authenticity and value of these documents and the sources of the cosmographical ideas of Columbus, followed by the various texts of the letter, with translations, annotations, several facimiles and also a map. Concludes that the Toscanelli letter is not authentic.

This is a translation of Vignaud's *La lettre et la carte de Toscanelli sur la route des Indes par l'ouest* (Paris: Leroux, 1901).

The conclusion is now regarded as hasty, primarily because Fernão Martins has proved to be a real person, a Portuguese cleric at the court of Rome. See SEM, #132, I, 56-57, note 9.

[530] **L. Gallois.** "Toscanelli et Christophe Colomb," *Annales de Géographie* (Paris), 11 (1902): 97-110.

A strong argument for the authenticity of the Toscanelli correspondence.

[531] **Norbert Sumien.** *La correspondance du savant Florentin Paolo del Pozzo Toscanelli avec Christophe Colomb.* Paris: , 1927. 113 pp.

Renounces his allegiance to Vignaud's position (see #529) and acknowledges that he thinks T's letter genuine.

[532] **Frederick A. Kirkpatrick.** "Toscanelli," *Hisp Amer Hist Rev*, 15 (1935): 493-95.

In answer to Vignaud (see #529) offers 4 reasons why T's 1474 letter should not be considered a forgery.

Mug 53

[533] **Charles E. Nowell.** "The Toscanelli Letters and Columbus," *Hisp Amer Hist Rev*, 17 (1937): 346-48.

Suggests that the authenticity of a correspondence between Toscanelli and CC is supported by João de Barros' statement that in

1484 when CC was interviewed by King John of Portugal the mariner was familiar with Marco Polo, and supported also the statements of Barros, Ruy de Pina, and García de Reserde that CC mentioned Cipangu in the interviews. Since "Cipangu" occurs only in Marco Polo, and since CC had not yet read Marco Polo in 1484, it is likely that the information came from the putative letter from Toscanelli.

[534] **Thomas Goldstein.** *Dawn of Modern Science from the Arabs to Leonardo Da Vinci.* Boston: Houghton Mifflin, 1980. 297 pp.

Pp. 15-29, a concise account of Toscanelli's contributions to the discovery of America.

c. The "Invention" of America

[535] **Edmundo O'Gorman.** *La idea del descubrimiento de América: historia de esa interpretación y crítica de sus fundamentos.* Mexico City: Centro de Estudios Filosóficos, 1951.

CC did not "discover" America, because to be ripe for discovery something has to be "fully predetermined and constituted." CC had no idea what he would find.

First rendition of the system of ideas completed in *The Invention of America* (Bloomington: Indiana Univ. Press, 1961); see #537.

[536] **Edmundo O'Gorman.** *La invención de América: el universalismo de la cultura de occidente.* Mexico City & Buenos Aires: Fondo de Cultura Económica, 1958. 132 pp.

A sequel to #535. America was invented, not discovered. CC set out to prove the existence of the Golden Chersonese (an unknown southern continent); Vespucci, to prove that all the lands to the west were parts of Asia. In the end the two men exchanged hypotheses. CC failed to transcend the archaic hypothesis he set out with, but V's theory contained empirical elements that led to the conception of a new continent.

Anticipates O'Gorman's *The Invention of America*, (1961), (#537); but the latter is an elaboration and extension, not a translation, of this book.

[537] **Edmundo O'Gorman.** *The Invention of America: An Inquiry into the Historical Nature of the New World and the Meaning of its History.* Bloomington: Indiana Univ. Press, 1961. 177 pp.

I. (1-48) Defines "discovery" as an act deriving both from the intent to find the thing discovered and the ability to recognize the significance of the thing discovered. Rejects CC as discoverer, since he meant to reach Asia and never knew he hadn't.

II. (49-69) By the 15th c the *Oikoumene*, that jointly spiritual, ethical, and physical order which contains human society, and the *orbis terrarum*, that world-island surrounded by sea which is the proper scope of man's activity, are seen as coextensive; but man's world is not recognized as a dynamic entity that can expand beyond the *orbis terrarum*.

III. (71-124) The Spanish monarchs (and Peter Martyr in his phrase *novus mundus*) were deliberately ambiguous as to whether the new land was Asia, thus supplying the concept of a new place that might or might not be part of the old *orbis terrarum*. Vespucci's *Lettera* (4 Sept 1504) provides the first written evidence of a conception of the new lands as a geographical entity. Waldseemüller's world map (1507) confirms the concept, and with the assignment of the name by the *Cosmographiae Introductio* (1507) America has been invented.

IV. (125-45) The fact that the *Cosmographiae* views the *orbis terrarum* as more than just one island signals a new view of man's habitat as being any place he can possess. Asia, Africa, and Europe now emerge as different parts of a whole. This implies that European civilization can and should be adapted to new circumstances. Spain tried adapting America to the European model and failed; the British colonists tried to adapt the model to the new lands, and succeeded fabulously.

[538] **Wilcomb E. Washburn.** "The Meaning of 'Discovery' in the Fifteenth and Sixteenth Centuries,", *Amer Hist Rev* 68 (1962-63): 1-21.

Reply to O'Gorman, #537. Argues that a more careful study of terms used in the age of discovery must precede effective translation of the discoverer's acts and intentions into historical truth. Examines "discover," "invent," "*terra firma*," "new world," "Asia," "*orbis terrarum*," "the Indies," and "continent," and on this evidence declares absurd O'Gorman's attempt to transfer the credit of discovery from CC to Vespucci. References to Vignaud's and Levillier's thesis-oriented scholarship.

[539] **David Beers Quinn.** "New Geographical Horizons: Literature," *First Images of America* (Berkeley: Univ. of California Press, 1976), 2: 653-58.

Traces the slow emergence of the concept of America as a separate entity in the writings (1) of CC, who viewed his discoveries as a part of Asia; (2) of Vespucci, who brought the new world to life, rendering it both novel and acceptable to the European reader; (3) of Peter Martyr, whose writings made the New World a fact of life; (4) of Cortés, whose narratives made the reader a spectator of the most striking episode in the Spanish explorations; and (5) of Oviedo, in whose work America had become assimilated into European concepts.

d. Controversial Maps

TOSCANELLI

[540] **Sebastiano Crinò.** *Come fu scoperta l'America: A proposito dell'identificazione della carta originale di Paolo dal Pozzo Toscanelli la cui copia servì di guida a Cristoforo Colombo per il viaggio verso il nuovo mondo.* Milano: Hoepli, 1943.

Infers that the planisphere of 1457 is a Toscanelli map made in Florence for the Castellani family, using an analysis of the handwriting and of the coat of arms, *inter alia.*

[541] **H. R. Wagner.** "Die Rekonstruktion der Toscanelli-Karte vom J. 1474, und die pseudo-Facsimilia des Behaim-Globus v. J. 1492," Vorstudien zur Geschichte der Kartographie III. *Nachrichten der K. Gesellschaft der Wissenschaften zu Gottingen,* Philolog.-historische Klasse, Nr. 3 (1894), 208-301. Rpt. *Acta Cartografica* (Amsterdam), 14 (1972): 352-461.

A flat projection (pp. 402-403) of the Western Ocean with Europe and Africa on the right margin, and Cathay, Mangi, and the island of Cippangu (among dozens of other islands) occupying the left 1/3 of the map. Antillia is an island running north-south halfway between the Canaries and Cippangu. *The map is not a representation of the Eurasian continent and not a planisphere like Fra Mauro's.*

THE GENOESE WORLD MAP OF 1457

[542] **Edward Luther Stevenson.** *Genoese World Map: 1457: Facsimile and Critical Text incorporating in Free Translation the Studies of Professor Theobald Fischer revised with the Addition of Copious Notes.* New York: American Geographical Society and Hispanic Society of America, 1912.

Critical text accompanies full-size color reproduction of the copy of the 1457 map in the Harvard Pusey Library Map Room. The original is in the Biblioteca Nazionale Centrale, Florence, Section Palatina No. 1, Medici Planisphere.

[543] **Sebastiano Crinò.** "La scoperta della carta originale de Paolo dal Pozzo Toscanelli che servì da guida a Cristoforo Colombo per il viaggio verso il nuovo mondo," *L'Universo* (Florence) (1941): 379-410.

Claims that the 1457 map is the famous Toscanelli map that encouraged and perhaps guided CC.

[544] **R. Biasutti.** "E stata ritrovata a Firenze la carta navigatoria di Paolo dal Pozzo Toscanelli?" *Riv Geog* (Florence), 48 (1941): 293-301.

The elliptical Florentine world map of 1457, attributed by Sebastiano Crinò (#543) to Toscanelli, is not inferior to that of Fra Mauro (1457), but is not consonant with the conceptions developed by Toscanelli and CC. It fuses classical and modern sources and represents developments that only prepare the way for CC's and Toscanelli's concepts.

[545] **Sebastiano Crinò.** "Ancora sul mappamondo del 1457 e sulla carta di . . . Toscanelli," *Riv Geog* (Florence) 49 (1942): 35-43.

Answer to #544 by R. Biasutti. The 1457 map is indeed in unison with the ideas of CC and Toscanelli, in spite of what B. says.

[546] **Alberto Magnaghi.** "Ancora intorno alla carta attribuita a Paolo dal Pozzo Toscanelli," *Riv Geog* (Florence), 49 (1942): 141-54.

Reasserts objections to Crinò's claims in #s 543 and 545 that the 1457 Florentine mappamondo is Toscanelli's and calls on Crinò to stick to the point in answering objections.

[547] **R. Biasutti.** "Il mappamundo del 1457 non é la carta navigatoria de . . . Toscanelli," *Riv Geog* (Florence), 49 (1942): 44-54.

Cf. #544. Makes his case more explicit, and again refuses to accept Sebastiano Crinò's assertion (#s 543, 545) that the 1457 Florentine world map is the *carta navigatoria* of Toscanelli.

HENRICUS MARTELLUS

[548] **Ilaria Luzzana Caraci.** "L'opera cartografica di Enrico Martello," *Riv Geog* (Florence), 83 (1976): 335-44.

A recent article by the Argentinian Pablo Gallez focuses attention on Henricus Martellus and his planisphere (ca. 1489). Gallez claims that one can see South America delineated in the 4th Asiatic peninsula. But cartographic and toponomastic analyis shows that the map derives largely from Marco Polo's experiences and the resulting conception of an undivided world. Gallez's view is unsupportable.

[549] **Arthur Davies.** "Behaim, Martellus, and Columbus," *Geog Jour* (London), 143 (1977): 451-59.

Asserts the following. The Nuremberg Globe of Martin Behaim, made in 1492, was almost identical with the Florentine Henricus Martellus's 1490 world map. Efforts to establish a common prototype in German cartography failed. A large world map signed by Henricus Martellus came to light ca. 1960 and was given to Yale Univ. It is an assembly of tracings on paper sheets of a world map completed by the Columbus brothers in 1490. Behaim may have seen the C. map in Lisbon in 1485. The Yale map is in Bartholomew C's hand. Martellus merely assembled the sheets to form his map.

AD

[550] **Ilaria Luzzana Caraci.** "Il planisfero de Enrico Martello della Yale University Library e i fratelli Colombo," *Riv Geog* (Florence), 85 (1978): 132-43.

Criticism of the hypothesis of Arthur Davies, #549, who, starting from the world map at Yale attributed to Henricus Martellus, maintained that CC did not use a planisphere similar to one by HM but rather that HM himself copied the world map in the British Museum from the map now at Yale, and that the Yale planisphere was made in Portugal by Bartholomew C.

In particular, the succession of the various codexes of the *Insularium illustratum* contradicts this hypothesis, and likewise a comparison of the representations of Africa and the Far East. LC feels that the main point to be clarified is how far the data set forth in the Martellus map coincide with CC's ideas.

FRA MAURO

[551] **G. R. Crone.** "Fra Mauro's Representation of the Indian Ocean and the Eastern Islands," *Studi Colombiani* (Genoa: SAGA, 1952), 3: 57-64.

Analyzes the designated portion of the map and concludes (1) that the extensive array of factual information is misaligned in ways that suggest the source to be not maps but perhaps fragments of a sailing directory; (2) that the map encourages sailing to the south and east to reach the Spice Islands, rather than to the west.

CHARLES DE LA RONCIERE

[552] **Charles de la Roncière.** *La Carte de Christophe Colomb.* Paris: Champion, 1924. 41 pp.

Cardboard case includes large folded reproduction of a map in the Bibliothèque Nationale that La Roncière attributes to CC. He makes these points: the map was made between 1488 and 1493; the mapmaker was Genoese; he copied passages from D'Ailly's *Imago Mundi*; the mapmaker had CC's own copy of *Imago Mundi* in his hands; the map was of the type that Bartholomew C traced; it was made under the very walls of Granada; it reveals that CC's goal was Antilia, or the Isles of the Seven Cities; CC took a more southerly route to Antilia than this map would show, and CC changed course to the SW to head for Cipangu, under the influence of M.A. Pinzón; CC had on board a map like this, not graduated in longitude and latitude. R. concludes with a discussion of CC's conception of the world, based on this map.

[553] **Albert Isnard.** "La carte prétendue de Christophe Colomb," *Revue des Questions Historiques* (Paris), 102 [2nd ser 6] (1925): 317-35.

Outlines Roncière's argument in #552, intended to demonstrate that a map in the Bibliothèque Nationale is by CC. Isnard refutes each point in turn, and rejects R's claim.

[554] **L. Gallois.** "Cartographie et géographie médiévales: une carte Colombienne," *Annales de Géographie* (Paris), 34 (1925): 193-209.

Argues in favor of La Roncière's conclusion (#552) that the map found in the Bibliothèque Nationale is by CC.

[555] **George E. Nunn.** "A Reported Map of Columbus," *Geog Rev* (NY), 15 (1925): 688-90.

Denies Roncière's contention (#552) that the map is earlier than 1492 and that it reflects CC's conceptions. It is simply a late 16th-century portolan chart.

[556] **Marcel Destombes.** "Une carte intéressant les études colombiennes conservée à Modène," *Studi Colombiani* (Genoa: SAGA, 1952), 2: 479-87.

Discusses a map of the European coast from Normandy to Naples, conceivably but not probably by Bartholomew C (and not by CC), which D. feels was made by the same hand as the famous map attributed to CC by La Roncière in 1925.

MARTIN BEHAIM

[557] **E. G. Ravenstein.** *Martin Behaim, his Life, and his Globe.* London: Philip, 1908. 122 pp.

A monumental publication drawing together a compendium of information about Behaim, who was CC's contemporary, possible acquaintance, and possible competitor as would-be discoverer of a western route to the Indies. Pp. 1-56, The Life, including "Behaim and Columbus," pp. 32-34; pp. 57-105, The Globe, consisting of a history, facsimiles, Behaim's sources, and an imposing "Nomenclature and Commentary." Inside end cover, a "Facsimile" of Behaim's globe, in color.

PIRI RE'IS

[558] **Paul E. Kahle.** "Impronte Colombiane in una carta turca del 1513," *Cultura,* 10 (1931): 775-85.

The "carta turca" is the Piri Re'is map (one of the few major discoveries about CC in the 20th century, along with Alicia B. Gould's research on CC's crews, #260).

Describes a fragmentary map, made ca. 1513 by a Turkish corsair named Piri Re'is, found by A. Deissman in Istanbul in 1929. Part of the map may reflect a map made by CC which fell into the Turk's hands when he captured 7 Spanish ships near Valencia in 1501. Legends on the map call CC a Genoese.

C. E. Nowell

[559] **Paul E. Kahle.** "A Lost Map of Columbus," *Geog Rev* (NY), 23 (1933): 621-38.

Piri Re'is, who drew his map in 1513, names as one source a map by "Colon-bo." Suggests that this map was one sent home to the sovereigns from Hispaniola after the discovery of Trinidad in 1498. Cuba is easternmost Asia, running north and south.

Mug 47

[560] **Paul E. Kahle.** *Die verschollene Columbus-Karte von 1498 in eine Türkischen Weltkarte von 1513.* Berlin and Leipzig: De Gruyter, 1933. 52 pp.

Discusses Piri Re'is, mariner and geographer, and his world map of 1513 with its mention of CC as a Genoese (see #558). Most of the monograph deals (1) with the details about the arrangement of the Antilles derived from the captured map of Columbus, and (2) the relationship of this map to the map of Toscanelli.

[561] *Piri Reis Haritasi.* Istanbul, 1935.

Reproduction of the Piri Re'is map with translation of text.

JUAN DE LA COSA

[562] **Cesáreo Fernández Duro.** "Mapamunde de Juan de la Cosa," *El Centenario* (Madrid), 1 (1892): 245-55.

A description and fold-out copy (in color), with separate, expanded details, of the map (dated 1500) acquired by the Spanish Ministro de Marina in 1853 from the Dutch baron de Walckenaer and displayed at the time in the Spanish Naval Museum. Mentions the possibility that the St. Christopher is a portrait of CC. Recounts important details, including detailed knowledge of Cabot's northern voyage, of Vicente Yañez Pinzón's 1500 discovery of Cape St. Augustine in Brazil, and most surprisingly (for a map dated 1500) of Cuba's status as an island.

Follows with an account of what was known at the time of Juan de la Cosa's career.

[563] **George E. Nunn.** *The Mappemonde of Juan de la Cosa.* Jenkinstown: Beans Library, 1934. 52 pp.

The surviving copy of the map in the National Museum at Madrid bears the date 1500; but this date contradicts the fact that Cuba is shown as an island, since the Ocampo voyage of 1508 was made to settle this issue. Also, the map's data on South America would have been available only *after* 1500. Concludes that the Madrid map is a copy, perhaps several times removed from La Cosa's original, made about 1508 and revised to accommodate the facts discovered after 1500.

Mug 78

[564] **G. Piersantelli.** *Lo schizzo cartografico di Colombo e il San Cristoforo nella carta del suo pilota.* Genoa: Civico Istituto Colombiano, 1955. 12 pp.

Argues that the St. Christopher that presides over Juan de la Cosa's Mappamundi is a representation of CC at the time of the first voyage, before success had changed him.

[565] **Fernando Royo Guardia.** "Don Cristóbal Colón, la insularidad de Cuba, y el mapa de Juan de la Cosa," *Rev Indias* (Madrid), 28 (1968): 433-73.

Argues (1) that Morison's difficulties in attempting to trace CC's route along the south coast of Cuba in 1494 are due to changes in the coastline since CC's day due to submersion, silting, and the action of hurricanes; (2) that CC at least suspected during the 1494 exploration that Cuba is an island; and (3) that Juan de la Cosa's map is correctly dated as 1500, since the details called in doubt could indeed have been known before 1500.

[566] **Roberto Barreiro Meiro.** "Algo sobre la carta de Juan de la Cosa," *Rev Gen Marina* (Madrid), 183 (1972): 3-8.

Attacks Morison's assertion that Juan de la Cosa's map dated 1500 was actually made after 1508. Cites a series of maps prior to 1508, including Waldseemüller's, in which the representation of Cuba as an island is influenced by the la Cosa map. Prints copies of the Cantino map of 1502 and the map of the Biblioteca Oliviarana de Pesaro (1505), both of which show Cuba as an island with a shape distinctively like that of la Cosa's Cuba. Cites Peter Martyr's assertion that Vicente Yañez Pinzón circumnavigated Cuba before 1499. Concludes that the inscription on la Cosa's map that it was made in Puerto de Santa María in 1500 is correct.

[567] **Arthur Davies.** "The Date of Juan de la Cosa's World Map and its Implication for American Discovery," *Geog Jour* (London), 142 (1976): 111-116.

Contends that the "Juan de la Cosa" map cannot be dated later than 1501 because of the large error in the latitude of coasts and islands in the Caribbean. Concedes G. R. Crone's conclusion that the map is not by La Cosa, and suggests that Vespucci may have helped construct it, and may have had it taken to Portugal.

WORLD MAP OF 1502

[568] **Paolo Revelli.** "Un cartografo genovese amico a Cristoforo Colombo: Nicolò Caveri ('Nicolas de Cauverio')," *Rendiconti della Classe di Scienze Morale, Storiche e Filologiche* (Genoa), Ser. 8, 2 (1947): 449-58.

Documented argument identifying the maker of a world map (ca. 1502) in the "Archives du Service Hydrographique de la Marina" in Paris.

WALDSEEMÜLLER

[569] **B. H. Soulsby.** "The First Map Containing the Name 'America,'" *Geog Jour* (London): 19 (1902): 201-209.

Argues that a world map containing the name "America," discovered by Henry N. Stevens in an imperfect copy of the

Strasburg *Ptolemy* of 1513 (1) is by Waldseemüller; (2) is entirely different from the world map usually found in the supplement of the 1513 Strasburg *Ptolemy,* and is the prototype of that map; and (3) is actually earlier than the Waldseemüller map bearing the name "America" that appears in the *Cosmographiae introductio* published in St. Dié on 25 Apr 1507, the book that first proposed that the New World should be called "America."

BARTHOLOMEW COLUMBUS

[570] **Roberto Almagiá.** "Intorno ad una carta di Bartolomeo Colombo (1513)," *Riv Geog* (Florence), 42 (1935): 29-33.

Hypothesizes that a map in one copy of Peter Martyr's 1511 *Decades,* showing the coastline of Florida two years before Ponce de León's first expedition to the North American mainland, is a map made ca. 1512-13 by Bartholomew C. for King Ferdinand. A. suggests that Peter Martyr inserted it in a 1513 reissue of the 1511 first edition of his first two *Decades.*

[571] **George E. Nunn.** "The Three Maplets Attributed to Bartholomew Columbus," *Imago Mundi* (Leiden), 9 (1952): 12-22.

The maplets, reproduced here, are in a collection of voyages known as *Alberico* in the Biblioteca Nazionale in Florence. Concludes that the maplets are not entirely the work of BC. The maplets constitute one map, and are an interpretation of the C. brothers' concept of world geography, based on the 56 2/3 mile value of an equatorial degree. They illustrate better than any other maps the C. brothers' concept of Mundus Novus joined to Asia, which they held after the 4th voyage and which BC testified to in the *Pleitos.*

e. Magnetic Declination

[572] **Timoteo Bertelli.** "La declinazione magnetica e la sua variazione nello spazio scoperta da Cristoforo Colombo," *Raccolta,* IV,2 (1892): 7-99.

Historical study of various suggestions that compass declination and its variation were known before CC, leading to the 13 conclusions (p. 79), all indicating that in this area at least CC was first.

[573] **Alberto Magnaghi.** "Incertezze e contrasti delle fonte tradizionali sulle osservazioni attribuite a Colombo intorno ai fenomeni della declinazione magnetica," *Boll Soc Geog Ital* (Rome), Ser. 6, 10 (1933): 595-641.

A contribution to the discussion of whether CC was the discoverer of magnetic declination and its western deviation, with a close examination of Bertelli's study, #572. M. feels strongly that a radical critical re-examination of some traditional sources must be undertaken (especially Ferdinand C's *Historie*) before one can evaluate with greater precision CC's attitudes and actions as a mariner.

[574] **A. Crichton Mitchell.** "The Discovery of the Magnetic Declination," *Terrestrial Magnetism and Atmospheric Electricity*, 42 (1937): 241-280.

Studies the references in CC's sources that concern magnetic declination, and infers (1) that an easterly declination had been observed in NW Europe before 1492; (2) that CC attributed his own observations to the wrong causes, unaware of a general variation of declination; and (3) that down to 1600 no original "observation, discovery, or instrumental improvement" can be associated with any specific person.

Mug 62

f. The Alexandrine Bulls of Demarcation

[575] **Edward Gaylord Bourne.** "The History and Determination of the Line of Demarcation Established by Pope Alexander VI, between the Spanish and Portuguese Fields of Discovery and Colonization," *Yale Review*, 1 (1892): 35-55.

Attempts a complete review of the subject beginning with Nicholas V's bull of 18 Jun 1452 to the resolution of the conflicting claims of Spain and Portugal in 1750. Besides establishing a "right of discovery," the bulls gave Portugal title to Brazil, gave rise to monopolistic European colonialism, and stimulated the progress of geodesy.

Mug 11

[576] **Samuel E. Dawson.** "The Line of Demarcation of Pope Alexander VI in A.D. 1493 and that of the Treaty of Tordesillas in A.D. 1494; with an Inquiry Concerning the Metrology of Ancient and Medieval Times," *Proceedings and Transactions of the Royal Society of Canada*, Ser. 2, 5 (1899): 467-546.

Discusses, among other things, ancient and medieval measures of length and the lines of demarcation. Prints 3 bulls of 1493, with

English translations, and Jaime Ferrer's 1495 discussion of how to find the line.

Mug 24

[577] **H. Van der Linden.** "Alexander VI and the Demarcation of the Maritime and Colonial Domains of Spain and Portugal, 1493-94," *Am Hist Rev*, 22 (1916-17): 1-20.

A study of the authorship of four bulls of 1493, all acts of papal sovereignty in favor of Spain, apparently authored by the pope himself. The idea of a line between Spanish and Portuguese activities is first found in the confirmation of CC's privileges on May 28. The bull of 26 Sept revoked exclusive privileges for Portugal in the Orient, the target of CC's forthcoming 2nd voyage.

Mug 98

[578] **E. G. Davenport.** *European Treaties bearing on the History of the United States and its Dependencies.*

Washington DC: Carnegie Institution, 1917-1937. 4 vols.

Vol. 1, pp. 56-100, docs 5-8, four of Alexander VI's bulls of demarcation, *Inter caetera I* (3 May 93), *Eximiae devotionis* (3 May 93), *Inter caetera II* (4 May 93), and *Dudum siquidem* (26 Sept 93). Doc 9, the Treaty of Tordesillas between Spain and Portugal, 7 Jun 94. Text, English translation, and bibliography for each. Note however that a 5th bull is also implicated, i.e., *Piis fidelium* (25 Jun 1493); cf. #581 and #586.

[579] **Charles E. Nowell.** "The Treaty of Tordesillas and the Diplomatic Background of American History," *Greater America: Essays in Honor of Herbert Eugene Bolton* (Berkeley: Univ. of California Press, 1945), pp. 1-18.

The treaty of Tordesillas was a diplomatic victory for John II of Portugal, who regained some of the ground lost in the four [sic] bulls of Alexander VI. The treaty also secured the route to India for da Gama's expedition by excluding Spanish vessels south of Cape Bojador.

Mug 74

[580] **Luis Weckmann.** *Las bulas Alejandrinas de 1493 y la teoría política del papado medieval: Estudio de la supremacía papal sobre islas, 1091-1493.* Mexico: Ist. de Historia, 1949.

Traces the growth of the "omni-insular" doctrine which, as part of public European law, Alexander VI applied in the bulls *Inter caetera I* & *II* in May and Jun 1493 and *Dudum siquidem* in Oct [sic] 1493, the bulls that established the Alexandrine line of division between islands to be explored and possessed by Spain and Portugal; and applied by Julius II in Jan 1506 in confirming the Treaty of Tordesillas of Jun 1494, which moved the initial Alexandrine line to a point 370 leagues west of the Cape Verde Islands.

The final chapter, pp. 229-61, details the application of the "omni-insular" doctrine to CC's discoveries. Concludes, i.a., that neither CC nor Alexander VI contemplated a continent as part of the negotiations.

[581] **A. García Gallo.** *Las bulas de Alejandro VI y el ordenamiento juridico de la expansion portuguesa y castellana en Africa y Indias.* Madrid: Instituto Nacional de Estudios Jurídicos, 1958. 369 pp.

An edition, bibliography, and elaborate historical and legal analysis of the 2 bulls *Inter caetera* of 3 and 4 May 1493; of *Piis fidelium* of 25 Jun 1493; and of *Dudum siquidem* of 26 Sept 1493. Follows the implications and consequences of the bulls through to describe the establishment of the juridical systems of the Portuguese and Spanish empires.

[582] **Arthur Davies.** "Columbus Divides the World," *Geog Jour* (London), 133 (1967): 337-44.

Proposes that the line of demarcation in the Treaty of Tordesillas (1494), resulting from the Alexandrine bulls of 1493, was the brainchild of CC, who conceived it as a means of keeping King John away from the new discoveries by giving the Portuguese a free hand in the Indian Ocean. Davies holds that CC had to find a way to prevent raids on his discoveries from the Portuguese, because his concessions in the Capitulations of Apr 1492 were not to be in effect until Spanish possession was assured, and Isabel did not intend to enforce possession by naval action. CC, Davies suggests, struck a deal with King John while visiting him in Portugal in early March—a deal that saved Spain from a naval struggle with Portugal, saved Portugal from a land invasion by Spain, and gave both nations a huge sphere (or hemisphere) of influence.

[583] *Tratado de Tordesillas y su proyección.* Valladolid: Seminario de História de América, 1973. 2 vols.

Essays based on presentations at the Primo Coloquio Luso-Español de Historia Ultramarina, Valladolid, Sept. 1972. Essays treat the background, the negotiations, the problems of application and enforcement, the effects on Africa, the Orient, and America, the consequent religious, anthropological, socioeconomic, and other problems, and the Spanish documentary support for the study of the treaty.

[584] **Miguel Battlori.** "The Papal Division of the World and Its Consequences," *First Images of America,* ed. Fredi Chiappelli (Berkeley: Univ. of Cal. Press, 1976), 1: 211-220.

Argues that Isabel, Ferdinand, and CC all considered the papal bulls of *Inter Caetera* of 3-4 May 1493 to be politically helpful but not essential to their goals of successfully claiming political jurisdiction for Spain of the lands discovered by CC.

[585] **Luis Weckmann.** "The Alexandrine Bulls of 1493: Pseudo-Asiatic Documents," *First Images of America,* ed. Fredi Chiappelli (Berkeley: Univ. of California Press, 1976), 1: 201-209.

Points out that since CC himself did not know he had discovered a new continent, the Alexandrine bulls (the basis of Portuguese claims to Brazil) cannot reasonably be viewed as a division of a continent (South America) whose existence was still unsuspected. It was, rather, a grant of the islands found by CC to Spain.

[586] **Lazzaro María de Bernardis.** "Le bolle Alessandrine, San Roberto Bellarmino, e la 'potestas indirecta in temporalibus,'" *Atti III Conv Internaz Stud Col* (Genoa: CIC, 1979), pp. 547-64.

The 5 Alexandrine bulls of 1493 *in re* CC drew their authority from the 'potestas directa' inherent in the rigid application of the theocratic system.

3. Biology

a. General

[587] **Manuel Colmeiro.** *Primeras noticias acerca de la vegetación americana.* Madrid, 1892. 59 pp.

Pp. 5-41, surveys the following early writings on the New World for botanical references: those of CC, Las Casas, Chanca, Peter Martyr, V. Y. Pinzón, Alonso de Ojeda, Hernan Cortés; and especially Oviedo. The remaining pages, 43-59, review the specifically botanical expeditions of the Spanish.

[588] **Alberico Benedicenti.** "Cristoforo Colombo e la medicina," *Studi Colombiani* (Genoa: SAGA, 1952), 3: 117-24.

Studies the impetus given to exploration in CC's time by the quest for medical remedies, and cites the centrality of Genoa's activity in this quest. Points out the initial contributions of coca and balsam, the peanut (this last one used in treating renal complaints), guaiaco, and tobacco.

[589] **Allesandro Brian.** "La fauna marina ai tempi di Cristoforo Colombo," *Studi Colombiani* (Genoa: SAGA, 1952), 3: 71-74.

Identifies, as sea-fauna referred to by CC, *tonni* (tuna), *remora* (reverso), and Sargassi algae or seaweed. The worms that perforated the ship *Viscaino* on the 4th voyage were *teredini*, molluscs that bore in wood.

[590] **Giovanni Pesce.** "I medici di bordo ai tempi di Cristoforo Colombo," *Studi Colombiani* (Genoa: SAGA, 1952), 3: 75-83.

Within the context of a series of documentary accounts of medicine as practiced at sea in the late Middle Ages, sets forth what is known of the medics on the 1st voyage, Maestro Alonzo of the *Santa María* and Maestro Juan of the *Pinta,* the latter presumably left at Navidad; and Diego Alvarez Chanca on the 2nd voyage.

[591] **Antonello Gerbi.** *La natura delle Indie nove. Da Cristoforo Colombo a Gonzalo Fernández de Oviedo.* Milan and Naples: Ricciardi, 1975. 631 pp.

English translation by Jeremy Moyle: *Nature in the New World: From Christopher Columbus to . . . Oviedo.* Pittsburgh: Univ. of Pittsburgh Press, 1986. 462 pp. Spanish tr. by Antonio Alatorre, Mexico: Fondo de Cultura Económica, 1978.

Annotation based on Moyle tr. CC treated as the first writer to describe plants, animals, and human beings of the western hemisphere (12-22); again in the section on Vespucci (35-44); and a third time (154-56) in the comprehensive biography of Oviedo that constitutes most of the book. Gerbi is frank about the self-serving slant in CC's observations on nature, but insists that CC is Oviedo's supreme hero. Gerbi exonerates Oviedo from the charge of being an anti-Columbian, a charge deriving from Ferdinand C's irritation with Oviedo for telling the "Portuguese pilot" story. See also #592.

[592] **Antonello Gerbi.** "The Earliest Accounts on the New World," *First Images of America* (Berkeley: Univ Cal Press, 1976), 1: 37-43.

Reviews the features and differences in the accounts of Columbus, Chanca, Cuneo, Vespucci, Peter Martyr, Oviedo, Enciso, Pigafetta, and Cortés. Reserves special praise for Vespucci and Oviedo as truly disinterested observers. Closely related to Gerbi's *La Natura delle Indie Nove,* #591.

[593] **Jonathan D. Sauer.** "Changing Perceptions and Exploitation of New World Plants in Europe, 1492-1800," *First Images of America* (Berkeley: Univ Cal Press, 1976), 2: 813-32.

CC and his early successors were unequipped to classify the various plants native to the New World. Eighteenth-century taxonomists constructed, from a bewildering range of observations, orderly classifications that intuitively expressed some evolutionary relationships, and this led to Darwin's critical recognition that species were evolved, not specially created. The most useful native American plants for Europeans were maize, tobacco, the potato, the tomato, the chili pepper, followed by the avocado, papaya, cacao, allspice, and chevimoyas.

[594] **Francisco Guerra.** "La epidemia americana de influenza en 1493," *Rev Indias* (Madrid), 45 (1985): 324-47.

The first Spanish-American epidemic, described as "gripe del cerdo," occurred in 1493, many years before the outbreak of smallpox in 1518. These conclusions derive from CC's and Dr. Chanca's remarks, and other information from Peter Martyr, Oviedo, Ferdinand C, and Herrera.

[595] **Francisco Guerra.** "El efecto demográfico de las epidemias tras el descubrimiento de América," *Rev Indias* (Madrid), 46 (1986): 41-58.

The profound social, political, and economic consequences of the infectious diseases exchanged between the Europeans and native Americans, e.g., the radical decline in American Indian population.

[596] **Francisco Guerra and C. Sánchez Teller.** "Las enfermedades de Colón," *Quinto Centenario* (Madrid), no. 11 (1986), pp. 17-34.

CC's infection and death by Reiter's syndrome, a disease due to the bacillus *shigella flexneri*, common in the tropics.

[597] **Pier Augusto Gemignani.** *La scoperta di Colombo e la medicina.* Genoa: Edizioni Culturali Internazionali Genova, 1988. 93 pp.

Monografie su temi colombiani, 3. A review of (1) what is known of the reciprocal effects of Old and New World diseases on the populations of the two worlds in the years following CC's first voyage, and (2) of the treatment of these diseases. Of special interest: the review of the literature on the genesis and spread of syphilis and related diseases like pinta and yaws, and the citation of Francesco Guerra's 1986 account of CC's infection by Reiter's syndrome, #596.

b. Syphilis

[598] **J. K. Proksch.** *Die Litterature über die venerischen Krankheiten.* Bonn, 1889-91. 3 vols.

Historiography of the issue.

[599] **Iwan Bloch.** *Der Ursprung der Syphilis.* Jena, 1901.

Argues for Columbian origin.

[600] **Karl M. Sudhoff.** "La supuesta introducción de la sífilis por la tripulación de Colón, en 1493," *Investigación y Progreso* (Madrid), 3, no. 9 (1929), pp. 65-67.

Argues against Columbian origin.

[601] **Richmond C. Holcomb.** *Who Gave the World Syphilis? The Haitian Myth.* New York: Froben, 1937. 189 pp.

Intro by C. S. Butler. Bib., pp. 177-89. Analyzes and criticizes Ruy Díaz de Isla's *Tractado contra el mal serpentino* (Seville, 1539). Concedes that Díaz's clinical description is unexcelled for his time, but discounts any suggestion that Díaz treated the disease in Barcelona in 1493 just after CC's arrival there with his Indian captives. Supersedes H's earlier treatments of the subject.

Mug 40

[602] **Emiliano Jos.** "El Centenario de Fernando Colón y la enfermedad de Martín Alonso," *Rev Indias,* 3, no. 7 (1942), pp. 96-101.

Concludes (p.100) that M.A. Pinzón returned from the West Indies in 1493 mortally infected with syphilis.

[603] **Samuel Eliot Morison.** "The Sinister Shepherd," *Admiral of the Ocean Sea* (Boston: Little, Brown, 1942), II, 193-218.

In #132. An extensive treatment. Concludes that CC's crews or his Indian captives introduced syphilis into Europe in 1493.

[604] **M. Lungonelli.** "Colombo e il morbo Gallico," *Bol Civico Ist Col,* (Genoa) 1, no. 2 (1953), pp. 51-64.

Although syphilis might have existed in Spain before the discovery (Peter Martyr refers to something like it in 1489), most scholars and sources conclude that it was brought to Europe from the New World, where it existed in a mild form.

Includes a 6-page bibliography and a chronology of the manifestations of syphilis from 1489 to 1539, when Ruy Díaz de Isla said that CC's return in 1493 began an epidemic in Barcelona.

[605] **Francisco Guerra.** "The Problem of Syphilis," *First Images of America,* ed. Fredi Chiappelli (Berkeley: Univ. of California Press, 1976), II, 845-51.

The close examination of the clinical observations in Ruy Díaz de Isla's *Tractado contra el mal serpentino* (1539) has enabled medical researchers to establish a calendar of the disease and its epidemiology. Concludes that venereal syphilis was brought back to Spain and Portugal by the discoverers; but the various forms of the disease all derive from an infection called *pinta,* which was contracted in its earliest forms from animals in Africa and carried thence to Siberia and America. From *pinta* developed endemic syphilis, venereal syphilis, and yaws. Probably none of these descendants of *pinta* was restricted to America prior to CC's voyage.

[606] **Earl J. Hamilton.** "What the New World Gave the Economy of the Old," *First Images of America* (Berkeley: Univ. Cal Press, 1976), 2: 853-84.

Appendix on syphilis (879-880) reviews the onset in Europe and concludes that the devastating European epidemic that began in the 1490's almost certainly came from infected persons in CC's returning ships in 1493.

[607] **Charles H. Talbot.** "America and the European Drug Trade," *First Images of America* (Berkeley: Univ. Cal Press, 1976), 2:833-44.

Although the direct influence of drugs found in the New World on European medicine was negligible except for the use of chinchona derivatives, the false claims made for heavily imported remedies led gradually to a scientific approach to therapeutics. T. uses the syphilis epidemic, attributed to effects of CC's 1st voyage, as a classic example: the enormous demand for the supposed cure, guaiacum, enriched the importers of this wood but ultimately led to experimentation with various cures and organized study of the results.

F. COLUMBUS AND HIS RELATIVES

1. General

[608] **Troy S. Floyd.** *The Columbian Dynasty in the Caribbean, 1492-1527.* Albuquerque: Univ. of New Mexico Press, 1973. 294 pp.

First half discusses the administration of CC and his brothers, embracing Cuba, Hispaniola, Puerto Rico, Jamaica, and the

Columbus claims on the mainland. Second half, rich in information on the Spaniards in the islands, discusses Diego's erratic administration and the developing exploitation of Hispaniola under his rule.

2. Forebears and Siblings (See also IV.B)

[609] **J. M. del Valle.** *Bartolomé Colón: Primer adelantado de Indias.* Madrid: Gran Capitán, 1946. 179 pp.

Not annotated, though based (to trust the bibliography) on primary sources. Full of "creative non-fiction," i.e., fiction introduced to fill out the story. Included here to indicate the gap in the scholarly record of CC's brother.

[610] **G. Balbis.** "Per la storia dei Colombo in Liguria nel secolo xv," *Atti III Conv Internaz Stud Col* (Genoa: CIC, 1979), pp. 219-34.

Guidelines for achieving direction and coherence in studying the Columbus family in Liguria in the 15th century.

[611] **Luis Fernández Martín.** "Un pleito de Bartolomé Colón relacionado con La Gomera," *Anuario de Estudios Atlánticos* (Madrid-Las Palmas), no. 29 (1983), pp. 15-37.

Attempts to establish (1) the motive for Bartholomew C's suit, which concerned BC's first voyage to the Indies in 1494, (2) the incidents in the suit, and (3) the reason for the long duration of the suit, i.e., 1500-1525.

Carmen Gómez Pérez

[612] **Antonia Heredia Herrera.** "Documentos colombinos en el archivo de la Diputación de Sevilla," *Archivo Hispalense* (Seville), 66, no. 203 (1983-84): 101-108.

Treats among other items a letter from Diego C and the will (1614) of Bartholomew C, which incidentally dates BC's death.

María Luisa Laviana Cuetos

[613] **Aldo Albonico.** "Bartolomeo Colombo, adelantado mayor de las Indias," *Pres Ital Andalu II* (Bologna: Cappelli, 1986), pp. 51-70.

Synthesizes what is known of Bartholomew C's life and career, beginning with a hitherto unexamined packet of testimony from the archives of Simanca containing testimony about BC's heroic conduct of his duties as Adelantado Mayor in Hispaniola up to the time when Bobadilla sent him home in chains along with CC, in 1500.

[614] **Gaetano Ferro.** "I luoghi di Colombo e della sua famiglia in Liguria," *Pres Ital Andalu II* (Bologna: Cappelli, 1986), pp. 135-42.

A chronological account of the places in Liguria with documented links to CC's family, i.e., the Colombos of Moconesi and the Fontanabuonas of Ponte di Cicagna, both settlements in a mountain valley of Liguria northeast of Genoa. On moving to the area of Genoa the families inhabited Quinto; houses in the Borgo di Santo Stefano at the edge of Genoa, near the Soprana gate; Savona, where CC's father moved his immediate family in the early 1470's; and the Borgo of Santo Stefano again, to which Domenico returned in the early 1480's. Incidental to these moves are various documented actions of the family members, concluding with the settlement by CC's son Diego of a claim against CC and his father Domenico.

3. Descendants

[615] **John Boyd Thacher.** "Arbor Consanguinitatis," Part X of *Christopher Columbus* (New York: Putnam, 1904), 3: 617-641.

In #81. A history of CC's descendants, with elaborate genealogical table (3.616).

[616] **Henry Vignaud.** *La maison d'Albe et les archives Colombiennes, avec un appendice sur les manuscrits que possédait Fernand Colomb et un tableau généalogique.* Paris: Soc. des Americanistes, 1904. 17 pp.

Traces the fortunes of the Columbus archives after the death of CC's grandson Luis Colón, through the complex descent of the noble houses that possessed them. The monograph is a celebration of the Duchess of Berwick, who located and published a considerable number of CC's letters and other documents in the decade or so beginning with the fourth centennial in 1892. See #s 5, 6, and 9.

[617] **Emiliano Jos.** "Investigaciones sobre la vida y obras iniciales de don Fernando Colón," *Anuario Estud Amer* (Seville), 1 (1944): 525-698.

Rpt. *Investigaciones sobre la vida y obras iniciales de don Fernando Colón.* Seville: Escuela de Estudios Hispano-Americanos, 1945. 164 pp. Ch. 1 reviews (1) the studies and bibliographies devoted to FC's book collection; (2) the sources and treatments of FC's life, in the writings of (a) CC himself, (b) F's friends, (c) prominent contemporary writers, (d) Argote de Molina and Fray Jerónimo Román, (e) 17th, 18th, 19th, and 20th-c historians. Ch. 2 traces FC's activity as a writer: (1) collaboration in CC's *Book of Prophecies*;

(2) *Los albores de la librería fernandina*; (3) the memorandum or agenda borne to Castile by Ferdinand in 1509; (4) FC's first two important works, *Colón de concordia* and *Forma de descubrir y poblar en las Indias*, possibly plagiarized from CC; (5) various writings suggesting FC's assiduity in winning the favor and entering the service of Charles V. Ch. 3 gives an account of various matters such as FC's legitimacy, his epitaph, the dates of his service as a page at court, documents attributed to FC, and others.

Finally, note that Jos attributes to Fernan Pérez de Oliva the first attack on FC's *Historie* as a disfigurement of CC's actual life. Cf. #111.

[618] **José Torre Revello.** "Don Hernando Colón: su vida, su biblioteca, sus obras." *Rev Hist América* (Mexico City), no. 19 (1945), pp. 1-59.

A straightforward account of what is known of Ferdinand C's life (Torre is aware of CC's 1497 mention of Diego and Ferdinand as legitimate sons, but not aware of the document (pub 1928) showing that the crown legitimatized FC; see Manzano, #622). A review of what is known of FC's great book collection, now mostly dispersed and mainly valuable for those of CC's books that bear the Admiral's marginal notes; and in addition an account of FC's life of CC, and a bibliography of FC's other works.

[619] **Otto Schoenrich.** *The Legacy of Columbus: The Historic Litigations Involving his Discoveries, his Will, his Family, and his Descendants.* Glendale CA: Arthur H. Clark, 1949-50. 2 vols.

Brief introductory chapter summarizing the whole 3 centuries of lawsuits ("Los Pleitos") over CC's discoveries, will, family, descendants, inheritance, titles, etc., 1508-1796, and the fortunes of the family since then. Describes the controversy and litigation beginning with the Capitulations of 1492, proceeding through the series of suits against the crown to determine the value of the entailed estate created by CC's wills, resulting in judgments in 1511, 1520, and 1525, and the completion of the original litigation in 1556. After the direct male line ended in 1578, litigation resumed among the descendants until the final judgment in 1796 when the victorious litigant took over CC's titles. His current successor bears CC's Spanish name, Cristóbal Colón.

Genealogical tables.

[620] **Rafael Nieto Cortadellas.** *Los descendientes de Cristóbal Colón. Obra genealógica.* Havana: Fernández y Cia, 1952. 485 pp.

Traces many lines of descent, but because there are no tables and only a very inadequate index, the work is difficult to use.

Martin Torodash

[621] **Emilio Rodriguez Demorizi.** *Familias hispanoamericanas*, Vol. 1. Ciudad Trujillo, 1959.

A genealogical study of families influential in colonizing America, including CC's. Extensive documentation, excellent index.

Martin Torodash

[622] **Juan Manzano Manzano.** *La legitimación de Hernando Colón.* Seville: Univ. of Seville, 1960. 22 pp.

Concludes (1) that a document published by Altolaguirre in "Algunos documentos inéditos relativos a don Cristóbal Colón," *Bol R Acad Hist* (Madrid), 92 (1928): 513-525, #38, proves that Ferdinand C had been legitimatized, and (2) that CC could not have married Beatriz Enríquez when he returned from the discovery because the laws of Spain did not permit a grandee to marry a person of low estate.

[623] **Antonio Rumeu de Armas.** *Hernando Colón, historiador del descubrimiento de América.* Madrid: Instituto de Cultura Hispánica, 1973. 454 pp.

An enormously detailed analysis of what is known of Ferdinand C as humanist, book collector, scholar, and historian, and an equally detailed analysis of FC's life of CC (the *Historie*) and its fortunes as a book. Rumeu asserts the existence of an anonymous biography alongside FC's original text, as well as later interpolations and augmentations.

[624] **Richard L. Garner and Donald C. Henderson.** *Columbus and Related Family Papers, 1451-1902: An Inventory of the Boal Collection.* Univ. Park PA: Pennsylvania State Univ. Press, 1974. 94 pp.

Penn State Univ. Studies, no. 37. Inventories archives of eleven family groups from whom the descendants of CC inherited papers. 3/4 of the papers concern Diego Santiago Colón, a tenth-generation descendant. 545 entries. Family tree, p. 5.

[625] **Luis Arranz Marquez.** "La noblesa colombina y sus relaciones con la Castellana," *Rev Indias* (Madrid), 35 (1975): 83-122.

Rpt. Madrid: Oviedo, 1976. 44 pp. Places within the context of the Catholic Monarchs' policy of controlling the nobility the steps in CC's battle to secure the privileges granted in the *Capitulations of Santa Fé* and the steps in Diego's decision to marry into the house of Alba rather than Medina Sidonia, a decision that enabled Diego to survive the ensuing struggle to preserve his privileges against the crown's hostility. Concludes by glancing at the degeneration of the Colóns' fortunes at the hands of the morally bankrupt Luis Colón.

[626] **Luis Arranz Marquez.** "La herencía colombina en los primeros proyectos de descubierta y colonización," *Rev Indias* (Madrid), 37, nos. 147-50 (1977): 425-69.

The Columbian cycle of discovery ended with CC's death. Diego's projects of discovery from 1509 to 1511 were entertained by King Ferdinand but then were not supported. In the years 1512-13, the crown definitively ceased consulting the governor of Hispaniola about new projects such as those of Juan Ponce de León.

[627] **Demetrio Ramos Pérez.** *Los Colón y sus pretensiones continentales: los planes sobre Norte América, Venezuela, Mexico y Perú.* Valladolid: Casa-Museo de Colón, 1977. 97 pp.

Cuadernos Colombinos no. 7. Studies the unsuccessful plans of CC, his son Diego, and grandson Luis to bring the continental discoveries under their domination and control.

[628] **F. Castellano.** "Domestici di Cristoforo e Diego Colombo," *Atti III Conv Internaz Stud Col* (Genoa: CIC, 1979), pp. 515-20.

Cites records of a number of CC's and Diego's domestic servants who attained positions of prominence, maintaining contact with Genoese merchants in Seville and with brokers in Seville and elsewhere in the south of Spain.

[629] **Luis Arranz Marquez.** *Don Diego Colón, Almirante, Virrey y Gobernador de las Indias.* Vol. 1. Madrid: Consejo Superior de Investigaciones Científicas, 1982. 392 pp.

A copiously documented biography, this volume terminating 1511. Pp. 161-392, 70 documents supporting the text.

[630] **Alberto Boscolo.** "Diego e Fernando Colombo: paggi alla corte dei Re Cattolici," *Temi Colombiani* (Genoa: ECIG, 1986), pp. 52-59.

Rpt. *Saggi su Cristoforo Colombo* (Rome: Bulzoni, 1986), pp. 51-60. Follows the career of CC's children at court. Shortly after 25 Sept 1493, the beginning of the 2nd voyage, Bartholomew C. took the children to court to be part of the entourage of Prince John, who like Diego C was 15 at the time. Ferdinand C. was 5. They received the same education as Prince John, from Peter Martyr and others, but also maintained ties with the Genoese in Spain, and with Beatriz Enríquez's relations, the Aranas. After Prince John's death, Diego and Ferdinand C became part of Queen Isabel's court.

In an aside, Boscolo shows that Ferdinand C, who was robbed of all his goods by a Turkish raider while on a voyage to Rome in 1512, may have lost a map that was the source of the information that Piri Re'is attributes to CC in his famous map.

[631] **Luis Fernández Martín.** *El Almirante Luis Colón y su familia en Valladolid (1554-1611).* Valladolid: Casa-Museo de Colón, 1986. 135 pp.

Cuadernos Colombinos no. 15. After a brief summary of what is known of Don Luis's life in Hispaniola, sets forth such facts as are known about the dissolute life of this "anti-hero" during his residence in Valladolid, Medina del Campo, and Simancas. The final chapter discusses Luis's descendants.

4. Supposed Relatives

[632] **Henry Harrisse.** *Les Colombo de France et d'Italie: Fameux marins du xve siècle 1461-1492.* Paris: Librairie Tross, 1874.

A monograph recounting what H. could find in the archives of Venice, Milan, and Paris about late 15th-c Mediterranean mariners named Colombo. The most important assertion for CC's biography is that Guillaume de Casenove, called Coullon, is the only one of these mariners who could have commanded a ship before 1485.

"Colombo Junior" or "Giorgio Griego" (George the Greek) and several others were much younger. Thus Harrisse clears the way for the later identification of Gillaume de Casenove as the commander of the squadron of French corsairs which attacked a Genoese convoy in Aug 1476 and which included the *Bechalla,* a ship wrecked in the fight and on which CC may have been a passenger. For further inf. see G. Pessagno, "Questioni Colombiane," 1926, #194.

G. COLUMBUS'S RELIGIOUS ZEAL, EVANGELISM, MYSTICISM

[633] **William K. Gillett and Charles R. Gillett.** "The Religious Motives of Christopher Columbus," *Papers of the American Society of Church History,* 4 (1892): 3-26.

The authors find no proof that before the 1st voyage CC suggested any religious dimensions to the Spanish monarchs for his Enterprise. Conclusion: the motive of conversion was subordinate, and late in developing.

Mug 32

[634] **P. F. Mandonnet.** *Les Dominicains et la découverte de L'Amérique.* Paris: P. Lethielleux, 1893. 255 pp.

Examines (1) the influence the cosmological ideas of the great Dominican doctors of the 14th c (Albert the Great and Aquinas) had on the scientific movement that prepared the way for the discovery of the New World and (2) the particular services rendered by the Frére Précheur Diego de Deza to CC.

[635] **John Leddy Phelan.** *The Millenial Kingdom of the Franciscans in the New World.* Berkeley: Univ. of Cal Press, 1956.

2nd ed. revised, 1970. Pp. 17-28, places CC within the Joachimite tradition.

[636] **Louis-André Vigneras.** "Saint Thomas, Apostle of America," *Hisp Amer Hist Rev,* 57 (1977): 82-90.

CC's identification of his discovery as "the Indies" promptly (in 1493) revived the tradition that the apostles Bartholomew and Thomas had evangelized India. In the discussions that followed, CC was compared with St. Thomas; and Fray Bernal Buyl (Buil,Boyl), the priest sent to evangelize the Indians on CC's 2nd voyage, was compared with St. Bartholomew. The belief that St. Thomas had evangelized the New World survived Magellan's proof that the Americas were not part of Asia.

[637] **Alain Milhou.** *Colón y su mentalidad mesianica en el ambiente franciscanista español.* Valladolid: Casa-Museo de Colón, 1983. 479 pp.

Cuadernos Colombinos no. 11. A systematic monograph addressing CC's religious mentality. Sets CC's messianism in the context of his typology, symbolism, and iconographic thinking. Attempts to place CC into religious trends in Spain and in Europe generally.

Delno C. West

[638] **Pauline Moffitt Watts.** "Prophecy and Discovery: On the Spiritual Origins of Christopher Columbus's 'Enterprise of the Indies,'" *Am.Hist.Rev.,* 90 (1985): 73-102.

Examines CC's image of himself as the fulfiller of certain apocalyptic prophecies in preparation for the Antichrist and the end of the world. Traces CC's image of the cosmos to Pius II's *Historia* and especially to d'Ailly's *Imago Mundi,* both owned and annotated by CC. Attributes the apocalyptic vision of CC's *Book of Prophecies* to d'Ailly's compendium of the eschatology of

Augustine, pseudo-Methodius, and Roger Bacon. Against this eschatology the biblical selections in CC's *B of P* manifest an obsession with the recovery of Jerusalem and the conquest and conversion of the heathen. In *B of P* and elsewhere, CC sees himself as fulfilling Seneca's prophecy of "new worlds" (*Medea* 376) and the medieval prophecy that the recapture of Jerusalem would come from Spain.

[639] **Leonard I. Sweet.** "Christopher Columbus and the Millennial Vision of the New World," *Cath Hist Rev* (Wash. DC), 72 (1986): 369-82, 715-716.

CC and the Spanish Catholics, not New England Puritans, introduced the millennial theme into the New World. CC saw himself as having taken the first of three steps soon to be completed which would lead to the dawn of the millennium. He believed he had discovered the Garden of Eden on the 3rd voyage and had opened up the hidden parts of the world, whose populations would soon be converted to Christianity. The other two steps, soon (in CC's view) to be achieved, were the recapture of Jerusalem and the conversion of the Jews. This view of millennialism as a central concern in the discovery of America helps explicate American history for several reasons: (1) this endemic character of millennialism in the European occupation of the New World helps explain the tenacity of the millennial ideal; (2) familiar American dichotomies such as egalitarianism-elitism and pluralism-homogeneity are somewhat clarified through a grasp of the millennial espousal of universal ideals accomplished through particularistic instruments; (3) millennialism helps explain why many Americans have been hostile to Catholic emigration while welcoming Jews: Catholics since the Reformation have been identified with Antichrist, while Jews brought into America might be more easily converted to Christianity.

H. COLUMBUS'S EDUCATION AND CULTURE

[640] **Berthold Laufer.** "Columbus and Cathay, and the Meaning of America to the Orientalist," *Journal of the American Oriental Society* (New Haven), 51 (1931): 87-103.

CC projected Chinese lore like dog-headed people and Amazons onto the aborigines of the New World. There was actual Chinese influence in America, filtered in via the land route down the Pacific coast, but this influence was slight.

Mug 50

[641] **Giuliano Raffo.** "Sulle postille di Colombo relativo alla storia romana," *Studi Colombiani* (Genoa: SAGA, 1952), 2: 69-75.

CC appears sufficiently steeped in Plutarch's *Lives* to repeat the latter's anachronisms in his postils.

[642] **Paolo Revelli.** "L'Italianità di Cristoforo Colombo," *Studi Colombiani* (Genoa: SAGA, 1952), 2: 9-38.

Reviews 3 questions: what is known (1) of CC's birthplace; (2) of the place where he received his fundamental culture and developed his fundamental personality; (3) of the complex of factors that define his nationality." Discusses birth (Genoa, Aug-Oct 1451); the Genoese character of CC's technical and practical knowledge of the sea; linguistic considerations; and the widespread 15th- and 16th-century recognition of CC as Genoese.

[643] **Emilio Rodríguez Demorizi.** "Colón y el refranero," *Studi Colombiani* (Genoa: SAGA, 1952), 2: 55-57.

Cites 44 Spanish proverbs and traditional sayings in the writings of CC, as evidence of the degree to which the Spanish language and culture had penetrated CC's mind.

[644] **Remedios Contreras Miguel.** "Conocimientos técnicos y científicos del descubridor del Nuevo Mundo," *Rev Indias* (Madrid), 39 (1979): 89-104.

Purpose: to analyze more thoroughly, within the limits of the title, works known to have been studied by CC before the discovery. Analyzes books used and annotated by CC that exist in the Biblioteca Colombina, Seville. These include works of D'Ailly, Silvio Piccolomini [Pius II], Marco Polo, and Zacuto, as well as CC's *Book of Prophecies*. Also analyzes at length the *Cosmographia* of Ptolemy in the collection at the Real Academia de la Historia. Cf. #s 77, 78, 164, and 481.

[644a] **Alfred Stückelberger.** "Kolumbus und die antiken Wissenschaften," *Archiv für Kulturgeschichte,* 69 (1987): 331-40.

CC's knowledge of the classics.

I. THE QUESTION OF JEWISH ORIGIN

[645] **D. Meyer Kayserling.** "Die Portugiesischen Entdeckung und Eroberungen in Beziehung zu den Juden," *Monatschrift für Geschichte und Wissenschaft des Judentums,* 7 (1857): 433-36.

Contains a preliminary treatment of the material in Kayserling's later *CC and the Participation of the Jews,* etc., #646.

[646] **D. Meyer Kayserling.** *Christopher Columbus and the Participation of the Jews in the Spanish and Portuguese Discoveries.* New York: Harmon, 1893. 189 pp.

Reviews, with careful attention to historical context, the prominent appearance of Jews and converted Jews in the story of CC's enterprise. A mine of information for Madariaga and others who have dealt with CC's putative Jewish antecedents. Develops the preliminary treatment in Kayserling's "Die Portugiesischen Entdeckung und Eroberungen in Beziehung zu den Juden," #645.

[647] **L. Modona.** *Gli ebrei e la scoperta dell'America.* Casale: Giovanni Pane, 1893. 92 pp.

Examines the following questions: (1) Did any Jews take part in the first voyage? (2) Was one of the women with whom CC lived a Jewess? (3) Did any Jew have direct or indirect influence on CC regarding his idea of traveling west to find new lands? Modona's answers: (1) no; (2) there is no evidence that he lived with a Jewess; (3) experienced Jews supplied effective counsel to the Portuguese maritime expeditions, and influential Spanish Jews provided key support for CC's expedition; but no one could enable even CC to determine ahead of time how to discover the New World.

[648] **Henry Vignaud.** "Columbus a Spaniard and a Jew?" *Amer Hist Rev,* 18 (1913): 505-512.

Rejects, on the basis of Genoese documents and the testimony of CC and his contemporaries, the claims that CC was a Spaniard or a Jew.

[649] **Jacob Wasserman.** *Columbus: der Don Quichote des Ozeans. Eine Biographie.* Berlin: S. Fischer, 1929.

English ed., Eric Sutton trans., *Columbus: Don Quixote of the Seas.* Boston: Little, Brown, 1930. 287 pp. A biography, without notes, and not closely in touch with the available facts (places CC's birth in 1436; 1451 had been reliably known as the date for more than 20 years), but which provided two strong incentives for Madariaga's book *Cristóbal Colón,* #650, by likening CC to Don Quixote (a likeness much emphasized by M.), and by pursuing the question of CC's possible Jewishness, which M. treats as a *cause celebre.*

W. emphasizes what he sees as CC's curious quixotry, and remarks in passing (pp. 150-51) a contrary strain in CC, a seeming Jewishness masked by secretiveness, suggesting that CC was perhaps a traitorous Jew. A full paragraph given to these non-quixotic Jewish traits (p. 151) seems to have caught Madariaga's eye and inspired the Don Quixote-Jewish split that animates his book. For further analysis of W's book, see H. P. Biggar, *Canadian Hist Rev* (Toronto), 12 (1931): 59-62.

[650] **Salvador de Madariaga.** *Christopher Columbus: being The Life of the Very Magnificent Lord Don Cristóbal Colón.* London: Hodder and Stoughton; New York: Oxford Univ. Press, 1939. 524 pp.

Spanish edition, Buenos Aires: Ed. Sudamericana, 1940. 657 pp. The main thesis of this thoroughly developed life is that CC was a Genoese Roman Catholic of Spanish Jewish extraction. Argues from personal traits and habits associated by M. with Jewishness.

Contains an interesting fantasy (Ch. 14) on CC's "poetic" nature which however sometimes seems simply to mean that he needed the love and services of his mistress Beatriz Enríquez de Harana.

Excellent portrayal of Isabel and Ferdinand throughout, and of Hernando de Talavera, chairman of the Spanish commission that studied CC's enterprise of the Indies (Ch. 13).

Caution: the Jewish argument is not effectively documented, and the Genoese of today claim that CC's "Jewish" traits and habits are traditionally those of Christian Genoese.

[651] **Armando Alvarez Pedroso.** "Cristóbal Colón no fue hebreo," *Rev Hist América* (Mexico City), no. 15 (1942), pp. 261-83.

Reviews the history of the question beginning with Aaron Goodrich's *A History of the Character and Achievements of the so-called CC* (1874), #124, and touching on the discussions of Maurice David in *Who Was Columbus* (NY 1933); W. F. McEntire (1925); Rafael Calzada (1925); Luis Ulloa (1928); and the simple claims of CC's Jewishness by Paredes (1903).

Since at this point Jewishness and CC's birthplace become intertwined, Alvarez treats the latter issue briefly, with special attention to García de la Riega's *Colón Español* (1914), #189, and the replies to claims of CC's Jewishness by Vignaud (1913), #648, and Altolaguirre, *Colón Español* (1923), #191. Finally A. turns to Madariaga, #650, at whom the article is chiefly aimed, and refutes M's arguments one by one.

[652] **Simon Wiesenthal.** *Segel der Hoffnung: Die Geheime Mission des Christoph Columbus.* Olten: Walter, 1972.

English trans. by R. & C. Winston, *Sails of Hope* (New York: Macmillan, 1973). Argues that CC, a man closely associated with Jews in Spain and perhaps a Jewish convert to Christianity, was secretly looking for lands settled by the lost tribes of Israel. He did not find these but he did find a land where the Jews, and the homeless and persecuted generally, could take refuge and prosper.

[653] **Michael Pollak.** "The Ethnic Background of Columbus: Inferences from a Genoese-Jewish Source, 1553-1557," *Rev Hist América*, no. 80 (1975), pp. 147-64.

CC's family "was not thought of either as Jewish or as converso by the citizenry of Genoa" during CC's lifetime.

Martin Torodash

[654] **Juan Gil.** "Colón y la Casa Santa," *Hist y Bibl Amer* (Seville), 21 (1977): 125-35.

Draws together a host of references in CC's writings suggesting that CC's obsession with Jerusalem, with Old Testament prophecies, and with finding the gold to enable the sovereigns to retake Jerusalem and rebuild the Temple (the "Casa Santa"), is part of a lifelong Judaic concern on CC's part with the coming of the Hebrew Messiah.

One of the most learned and startling of all treatments of CC's possible Jewishness.

J. CANONIZATION

[655] **Antoine François Félix Roselly de Lorgues.** *Christophe Colomb: histoire de sa vie et de ses voyages.* Paris: Didier, 1856. 2 vols.

The biography that led to the 19th century movement to canonize CC. Indulges in extensive "creative non-fiction" such as in the treatment of Beatriz Enríquez de Harana (CC's mistress and the mother of Ferdinand C) as a woman of noble family whom CC married and who rejoiced happily in the glory that accrued to him (1: 172-76). Generally, the book ignores all the unattractive aspects of CC in order to make him a saint. See Sanguineti, #656.

[656] **Angelo Sanguineti.** *La canonizzazione di Cristoforo Colombo.* Genoa: Sordo-muti, 1875. 18 pp.

Classic refutation of the arguments for canonizing CC intitiated by Roselly de Lorgues #655. Demolishes Roselly's argument that CC had married Beatriz Enríquez de Harana by showing misreadings and mistranslations of key texts and by advancing as a proof of his own arguments documents that make clear the illegitimacy of Ferdinand C. Sanguineti's theme is that Roselly lets sentiment control his argument and consequently ignores the patent facts of the matter. See also #141.

[657] **Giovanni Odoardi.** "Il processo di beatificazione di Cristoforo Colombo," *Studi Colombiani* (Genoa: SAGA, 1952), 3: 261-72.

Reviews sympathetically the movement, smiled on by the popes Pius IX and Leo XIII, and led by Roselly de Lorgues, to achieve the beatification and canonization of CC. Describes without prejudice the stumbling block that has so far prevented the movement from achieving success, i.e., Las Casas' assertion that Ferdinand C was CC's "natural son" by Beatriz Enríquez de Harana. Ends with a wistful further call for an objective, scholarly study of CC's life that would either establish or demolish the case for beatification.

[658] **Osvaldo Chiareno.** "La religiosità di Cristoforo Colombo e le polemiche sui tentativi per la sua canonizzazione," *La lingua di Colombo e altri scritti di Americanisti* (Genoa: DiStefano, 1988), pp. 27-50.

Reviews the extensive evidence of CC's enthusiastic religiousness, concluding with his acknowledgment at the end of his guilt over his treatment of Beatriz Enríquez de Harana; and rehearses briefly the abortive 19th-century campaign by Roselly de Lorgues to have CC canonized.

K. COLUMBUS'S NAVIGATION

[659] **Earl of Dunraven.** "Note on the Navigation of Columbus's First Voyage," in Filson Young, *Christopher Columbus and the New World of his Discovery*, 3rd ed. (New York: Henry Holt, 1912), pp. 399-422.

An elaborate analysis of the evidence in the *Journal*, including CC's ideas of the globe, his instruments, observations for latitude and longitude, course and distances, landfall, and dead-reckoning. Throws light on many passages in CC's narratives, e.g., the passage in the *Journal* for 30 Sept 1492, which interprets the "Guards" as Beta and Gamma in Ursa Major and explains the use mariners made of them.

[660] **Johan Menander.** "The Navigation of Columbus," *Proceedings of the U. S. Naval Institute* (Annapolis): 52 (1926): 665-73.

Describes the astrolabe, cross staff, and quadrant used by CC. CC's four recorded observations of latitude were double what they should have been, probably because Las Casas added CC's readings of both sides of the astrolabe. CC reckoned his distance to San Salvador at 350 miles more than the actual distance.

Mug 61

[661] **Jean Baptiste Charcot.** *Christophe Colomb vu par un marin.* Paris: Flammarion, 1928. 320 pp.

Preface by Paul Chack. A chronological account of CC's life, emphasizing the voyages in a way that anticipates Morison's *AOS*, #132. No attempt at the exhaustive documentation of *AOS*, but close attention to nautical details, leading to such observations as that CC's ships on the 1st voyage were in excellent condition and that M.A. Pinzón could not be solely responsible for the maritime success of the 1st voyage, since CC was complete master of his craft as seaman and dead-reckoner.

[662] **Alberto Magnaghi.** "I presunti errori che vengono attribuiti a Colombo nella determinazione delle latitudini," *Boll Soc Geog Ital* (Rome), 65 (1928): 459-94, 553-82.

Rpt. Rome: R. Soc. Geog. Ital., 1928. 67 pp. Scholars from Humboldt on have denigrated CC's knowledge of navigation even in those areas where the requisite information was already widespread in Italy and elsewhere. The errors attributed to CC, however, generally derive from (1) documents misunderstood or misread by modern scholars; (2) restrictions placed on CC (e.g., by the Spanish sovereigns) that prevented him from furnishing correct evidence; and (3) documents falsified by hands other than CC's after they left his hands. M. uses CC's presumed errors in latitude to illustrate this situation.

[663] **J. A. Williamson.** "The Early Falsification of West Indian Latitudes," *Geog Jour* (London), 75 (1930): 263-65.

A review of Alberto Magnaghi's article "I presunti errori, etc.," #662. Rejects Magnaghi's claim that CC's and La Cosa's errors of 21 deg. in the latitude of the West Indies were made to deceive the Portuguese. CC's were simply errors; La Cosa's were intended to fool the English, who had reached Newfoundland, and might be inclined to intrude on Spanish discoveries.

[664] **Alberto Magnaghi.** "Ancora dei pretesi errori di Colombo nella determinazione delle latitudine," *Boll Soc Geog Ital* (Rome), Ser. 6, 7-8 (1930): 497-515.

Rpt. Rome: R. Soc. Geog. Ital., 1930. 21 pp. Further elucidates, vis-à-vis J. A. Williamson (#663, previous item), M's position in "I Presunti Errori" (#662) that some of CC's gross misrepresentations of latitude were dictated by royal Spanish policy respecting Portuguese rights in the Atlantic.

[665] **A. Fontoura da Costa.** *A Marinharia dos Descobrimentos.* Lisbon: Geral das Colónias, 1933. 532 pp.

2nd ed., augmented, 1939. Elaborately detailed account of marine instruments, techniques, maps, and practices from the time of Henry the Navigator to 1700. For references to CC, Bartholomew C, and Ferdinand C, see Index, p. 515.

[666] **John W. McElroy.** "The Ocean Navigation of Columbus on his First Voyage," *American Neptune*, 1 (1941): 209-240.

A study based on CC's *Journal*, with computation of the effect of various perceived factors, to establish a daily position for the *Santa María*. This is the basis of Ernest Raisz's plot of the first voyage published by Morison in *AOS*, #132.

[667] **Samuel Eliot Morison.** "Columbus and Polaris," *American Neptune* (Salem MA), 1 (1941): 6-25, 123-37.

A technical study of CC's celestial observations for navigational purposes as recorded in the *Journal* and elsewhere. CC's errors of latitude, as in doubling the latitude of NW Cuba, were due to his mistaking another star for Polaris. CC's navigational genius lay in his dead-reckoning; when he had been to a place, he could sail to it again.

Mug 64

[668] **Samuel Eliot Morison.** "Columbus as a Navigator," *Studi Colombiani* (Genoa: SAGA, 1952), 2: 39-48.

CC, not skilled even in the navigational instruments available to him, was a consummate navigator by virtue of his uncanny skill at dead-reckoning and his gift for managing seamen. He trained the great Spanish navigators of his generation.

[669] **H. Winter.** "Bemerkungen zur Navigation von Kolumbus und der seiner Zeit," *Studi Colombiani* (Genoa: SAGA, 1952), 2: 49-54.

Whatever CC's shortcomings with his instruments, as in his misreading of latitude, he was an excellent observer and a great navigator.

[670] **E. G. R. Taylor.** "The Navigation Manual of Columbus," *Journal of the Institute of Navigation* (London), 5 (1952): 42-54.

Rpt. with facing-page Italian trans., *Boll Civ Ist Col* (Genoa), no. 1 (1953), pp. 32-45. Based on Fontoura da Costa's *A Marinharia dos Descobrimentos* (#665), which describes the evidence for early sailing methods, and on Joaquim Bensaude's reprints of rare documents for the Portuguese government, especially the *Repertorio dos tempos* of Valentim Fernández (1st ed. 1518). T. seeks to establish (1) that in an *Addendum* to the *Repertorio* we have extracts from an archaic navigating manual, and (2) that CC had studied navigational rules and instructions just such as the *Addendum* contains, but because of his deficient math he had not understood them.

[671] **Samuel Eliot Morison.** *Christopher Columbus, Mariner.* Boston: Little, Brown, 1955. 224 pp.

Maps by Erwin Raisz. A straightforward narrative of CC's life, based on the material in *AOS* (#132), but emphasizing the author's own conclusions on controversial matters. Appendix contains a new translation of the *Letter to Santangel* of 4 Mar 1493 recounting the 1st voyage.

[672] **Juan García Frias.** "Colón y la náutica en el siglo XVI," *Rev Gen Marina* (Madrid): 187 (1974): 297-313.

CC initiated the change from navigation by intuition to navigation by instrument. Surveys the initial steps in instrumental navigation beginning with CC's 1st voyage.

[673] **Rolando A. Laguardia Trias.** *El Enigma de las latitudes de Colón.* Valladolid: Casa-Museo de Colón, 1974. 67 pp.

Cuadernos Colombinos no. 4. Hypothesizes an explanation for CC's progress from completely erroneous readings of latitude (e.g., El Mina) to double readings (Cuba, 42 instead of 21 deg) to correct (Jamaica, 18 deg.)

[674] **Osvaldo Baldacci.** "Tecnica nautica fra medio evo ed età moderna," *Atti III Conv Internaz Stud Col* (Genoa: CIC, 1979), pp. 65-90.

A reflection on the concerted effort among mariners from medieval to modern times to make their gradually improving instruments serve the expanding purposes of ocean navigation.

[675] **James E. Kelley, Jr.** "The Navigation of Columbus on His First Voyage to America," *Columbus and his world,* ed. Donald T. Gerace, (Ft. Lauderdale FL: CCFL, 1987), pp. 121-40.

Summarizes the results of a computer simulation of C's navigation during the first voyage, based on course and distance data in the *Journal* and on other sources of information on 15th-c cartography and navigation. Discusses, i.a., CC's mile, his "double-accounting" of distance, and his "land league."

JEK

[676] **Georges A. Charlier.** *Étude complète de la navigation et de l'itinéraire de Cristóbal Colón lors de son voyage de découverte de l'Amérique.* Vol. 1. Liége, Belgium: Charlier, 1988. 130 pp.

Charlier, a marine historian and blue-water sailor, uses CC's observations as found in the *Journal* and in Ferdinand C's *Historie.* Strives to make sense of conflicting evidence and incomplete records, and produces an extremely thorough study of CC's navigation. Does not argue for a particular landfall, but highlights the key issues that must be resolved in order to understand how CC navigated.

Philip Richardson

L. PORTRAITS

[677] **James Davis Butler.** *Portraits of Columbus: A Monograph.* Madison WI: privately published, 1883. 23 pp.

Address delivered to the Wisconsin Historical Society, 1882, largely devoted to the Giovian portrait of CC and those related to it. Concludes with a proposal for immortalizing members of Congress by standing them in a certain mineral spring in Yellowstone Park which will quickly impregnate them with minerals and turn them to stone, thus producing statues of them all.

[678] **William Eleroy Curtis.** *Christopher Columbus: His Portraits and his Monuments. A Descriptive Catalogue.* Chicago: Lowdermilk, 1893. 72 pp.

Part 1. One of the best treatments of the portraits, with recognizable photo or sketch of each item. Descriptions of 73 portraits—paintings, engravings, and woodcuts. Part 2. Descriptions of monuments: busts, statues, columns, buildings, shields, medals, with photos and sketches.

This information on portraits also appears in the *Report of the United States Commission to the Columbian Historical Exposition at Madrid, 1892-93* (Washington DC: Govt Printing Office, 1895), pp. 218-57.

[679] **Charles P. Daly.** "Have We a Portrait of Columbus?" *Jour Amer Geog Soc* (NY), 25 (1893): 1-63.

Argues for the authenticity of the Stimmer woodcut (1576) of the "Giovian portrait," the Altissimo copy (Uffizi, Florence), and the Capriolo engraving (1596).

Mug 21

[680] **Juan Pérez de Guzman.** "Retrato di Cristóbal Colón, descubridor del nuevo mundo," *El Centenario* (Madrid), 3 (1893): 415-26.

Argues for the authenticity of the portrait of CC in the Biblioteca Nacional in Madrid, with script "COLVMBVS LYGVR NOVI ORBIS REPTOR" across the top.

[681] **Nestor Ponce de León.** *The Columbus Gallery: The "Discovery of the New World" as Represented in Portraits, Monuments, Medals and Paintings. Historical Description.* New York: Ponce de Leon, 1893. 178 pp.

A descriptive and historical account. 1. Portraits, in 3 sections: a. pictures and engravings possibly taken from the life, and copies of them. These are traced to the no longer extant "Giovo portrait" through the copy in the Uffizi gallery and the Tobias Stimmer woodcut published 1575. Ponce considers the De Orchi copy (now in the Giovo museum in Como, Italy) to be the best extant representation of CC. The most influential of the Jovius descendants is the Capriolo engraving, pub. 1606, which gave rise to a large number of recognizably similar blond portraits of CC, more Germanic in facial type. b. Other portraits based on the descriptions of CC left by his contemporaries. c. *Imaginary* pictures and engravings.

[682] **Achille Neri.** "I ritratti di Cristoforo Colombo," *Raccolta*, II.3 (1894): 249-80.

Reviews the various portraits reputed to picture CC, and concludes there are only two with any likelihood of authenticity, i.e., the [De Orchi] painting in the Giovo Museum, Como, Italy, and the Capriolo engraving. These are the archetypes of the two traditions that may be authentic.

Includes 30 plates with reproductions of some 40 portraits reputed to represent CC. Bibliography, pp. 275-85, of 47 items treating CC's portraits.

[683] **John Boyd Thacher.** Section on portraits in Part VIII, "Personality," *Christopher Columbus* (New York: Putnam, 1903-4), 3: 1-79.

In #81. Classifies portraits of CC into 3 types: (1) Jovian type (including both of the most popular portraits, the De Orchi and Capriolo); (2) De Bry type, and (c) bearded type. A very thorough essay, including reproductions (with late 19th-century limitations) of 40 separate portraits and copies.

[684] **Armando Alvarez Pedroso.** "El verdadero retrato de Cristóbal Colón," *Studi Colombiani* (Genoa: SAGA, 1952), 3: 25-29.

Invokes the evidence of the bones in the coffin at the Santo Domingo cathedral to support the argument that the [De Orchi] portrait of CC in the Giovo Museum, Como, Italy, is an authentic likeness.

[685] **George Kish.** "The Caprarola Portrait of Columbus," *Geog Jour* (London), 120 (1954): 483-84.

Discusses the portrait of CC in the Palace of Caprarola, summer residence of the Farnese family in the Cimini hills north of Rome. Kish assigns the painting (not mentioned in W.E. Curtis's catalogue, #678) to the period 1559-74. It falls within the group labeled the "Giovo type" by Curtis. Kish thinks however that it is not based on the painting in the Archbishop's gallery at Como but on an earlier, now lost, painting done from the life, or on tradition, deriving from assertions of CC's contemporaries, or both.

M. MORTAL REMAINS

[686] **Real Academia de la Historia.** *Los Restos de Colón. Informe de la Real Academia de la Historia.* Madrid: Ministerio de Fomento, 1879. 197 pp.

Reviews the available evidence and concludes that CC's body was indeed in the cathedral at Santo Domingo until the transfer to Havana in 1795, and that at the time of this report CC's remains are in the cathedral at Havana.

[687] **John Boyd Thacher.** "Los Restos," Part VIII of *Christopher Columbus* (New York: Putnam, 1903-4), 3: 491-613.

In #81. An enormously detailed examination of the evidence, concluding that CC's body remains in the Cathedral at Santo Domingo. Photographic reproduction of the major evidence, including 21 photos.

[688] **E. Tejera.** "Acta de la entrega y depósito del cuerpo de D. Cristóbal Colón en el Monasterio de Santa María de las Cuevas de Sevilla," *Clio* (Santo Domingo), Jul-Aug 1933, pp. 94-96.

Prints a transcript of the Wednesday, 11 Apr 1509 order of Johan Rodríguez for the delivery and deposit of CC's body in the monastery, at the request of Diego C. Establishes that the date 11 Apr 1509 must be correct, because 11 April does not come on Wednesday in any other possible year.

[689] **F. Llaverías.** *Cristóbal Colón: El hallazgo de sus restos en Santo Domingo.* Havana: Fernández y Cía, 1939. 25 pp.

English edition also, same date and publisher. The discovery of CC's remains in Santo Domingo.

[690] **Antonio Ballesteros Beretta.** "Los Restos de Colón," *Bol R Acad Hist* (Madrid), 120 (1947): 7-49.

Reviews the evidence and concludes that no convincing argument can be made to support the thesis that CC is still buried in the Santo Domingo Cathedral. Votes for the remains in the Seville Cathedral as the authentic remains of CC.

[691] **Armando Alvarez Pedroso.** "Los restos mortales del descubridor de América don Cristóbal Colón," *Studi Colombiani* (Genoa: SAGA, 1952), 3: 15-23.

A summary of the historical events and physical evidence leading to the conclusions that CC's remains are still buried in the Cathedral at Santo Domingo and that the remains now buried in CC's monument in the cathedral at Seville are those of Diego C. Refers reader to the more complete account in A's *CC: biografía del descubridor* (#476), Ch. 26.

[692] **Cristóbal Bermúdez Plata.** "Los Restos de Colón," *Anuario Estud Amer* (Seville), 8 (1951): 1-11.

Summarizes the controversies over the location of CC's remains, including account of the transfer of a body (CC's? Diego C's?) from Santo Domingo to Havana in the 18th c, and then to Seville in 1898-99.

Martin Torodash

[693] **Manuel Giménez Fernández.** "Los Restos de Cristóbal Colón en Sevilla," *Anuario de Estudios Americanos* (Seville), 10 (1953): 1-170.

Argues that CC's remains do not rest either in the cathedral at Santo Domingo or in the Cathedral of Seville because they were never moved to Santo Domingo in the first place. Argument delivered in staggering detail, 170 pp. long.

[694] **Fernando Arturo Garrido.** "Vicisitudes del muerto inmortal," *El Faro a Colón* (Santo Domingo), 9: 20 (Jan-April 1958), pp. 81-116.

Abstract, documented, of the several divergent published theories on the last resting place of CC's remains. Concludes that the authentic remains are in the Santo Domingo cathedral.

Martin Torodash

[695] **Cipriano de Utrero.** "Los restos de Colón," *El Faro a Colón* (Ciudad Trujillo), 9: 20 (1958): 7-80.

Eloquent and persuasive defense of Santo Domingo as current resting place of CC's remains. Based on research in the Archive of the Indies, Seville. Bibliography.

Martin Torodash

[696] **Carlos Nouel.** "Carta del Señor Don Carlos Nouel," *El Faro a Colón*, 10: 24 (May-Aug 1959), 141-143.

Reprints Nouel's letter of 1878 asserting the authenticity of the remains in the cathedral at Santo Domingo.

Martin Torodash

N. COLUMBUS'S SHIPS

1. General

[697] **Enrico Alberto D'Albertis.** *Le costruzioni navali e l'arte della navigazione al tempo di Cristoforo Colombo. Raccolta*, IV.1. Rome: Ministero della Pubblica Istruzione, 1893. 240 pp.

An elaborate technical study. Chapters: 1. Medieval ships. 2. CC's ships *Santa María, Pinta, and Niña*. 3. Medieval cartography. 4. Daily reckoning of course and distance; altitude navigation; instruments. 5. Analysis of 1st voyage; Watlings as Guanahani; declination of compass; tables; date of CC's *Letter to Santangel*.

[698] **Cesáreo Fernández Duro.** "Armamento de las carabelas de Colón," *El Centenario* (Madrid), 1 (1892): 197-207.

Reviews what little direct evidence there is in documentary accounts of CC's voyages, and supplements this with what is known of the armament of other Spanish ships of the time.

[699] **Cesáreo Fernández Duro.** "La vida en las carabelas de Colón," *El Centenario* (Madrid), 3 (1893): 166-80.

A detailed discussion of what can be known about life aboard caravels in CC's day. Strongly anticipates the similar chapters in Morison's *AOS*, including the words of songs sung on board.

[700] **Rafael Monleón Torres.** "Las carabelas de Colón," *El Centenario* (Madrid), 1 (1892): 53-61, 119-128.

Part 1 studies the etymology of "carabela" and makes these points: (1) the term began to be used in Spain for ships like CC's in the middle of the 15th c; (2) they perhaps originated in Portugal; (3) their form varied, and the name designated the type of service, i.e., exploration, not cargo moving; (4) they were fast-sailing and stoutly built, with relatively high castles fore and aft, 3-masted, with square sails on the foremast and mainmast and lateen on the mizzen. *Santa María* presumably displaced 180-220 metric tons, the others less. Part 2 provides as detailed an analysis of the ships as M considers possible, and concludes with his recommendations for specifications of the 4th centenary Spanish model of the *Santa María*.

[701] **Heinrich Winter.** *Die Kolumbusschiffe von 1492.* Madgeburg, 1944. 54 pp.

2nd ed., Rostock: Hinsdorff, 1960. 54 pp.

Italian ed., D. Curti, tr. & ed., *Le navi de Colombo* (Milan: Mursia, 1972). 67 pp.

Discusses the construction of the ships.

Martin Torodash

[702] **Björn Landström.** *Skeppet.* Stockholm: Bokförlaget, 1961.

English ed. *The Ship: An Illustrated History,* tr. Michael Phillips (London: Allen & Unwin, 1961), 309 pp. Excellent illustrations. Section on CC parallels the treatment of ships in #135.

[703] **José María Martínez-Hidalgo.** *Columbus's Ships,* ed. Howard I. Chapelle. Barre MA: Barre Publishers, 1966. 123 pp.

Span. ed. *Las Naves de Colón* (Barcelona: Cadi, 1969), 230 pp. Thesis: *Santa María* was a não; *Niña* and *Pinta,* caravels. Reconstructs the *Santa María,* though not from 15th-century evidence. Lacks footnotes.

[704] **Carlos Etayo Elizondo.** *Naos y caravelas de los descubrimientos y las naves de Colón.* Pamplona: Aralar, 1971. 262 pp.

A detailed study (1) of the evolution of marine construction from the 13th c; (2) of the bases available for historical reconstruction of the ships in CC's 1492 fleet; (3) of the characteristics of the 3 ships in the fleet.

Pp. 217-241, detailed critical judgment of a series of attempts to reconstruct the ships, including those of the Comisión Española (1892), D'Albertis (1892-4), Guillén (1927-29), R.C. Anderson (1930), the author's own (1962), and M. Hidalgo (1969).

[705] **Carlos Etayo Elizondo.** "Las naves del descubrimiento," *Rev Gen Marina* (Madrid), 206 (1984): 711-721.

A protest against the plans of the Spanish 5th Centennial Commission for constructing a new model of the *Santa María.* Etayo presents what he considers adequate bases for the construction, derived from a study of the documents.

[706] **Eugene Lyon.** "15th-Century Manuscript Yields First Look at *Niña,*" *National Geographic* (Wash. DC), 170, no. 5 (1986): 600-605.

On the basis of a remark of Pedro Frances, in the ms. *Libro de Armadas* (1498), that the *Niña* when he received it for CC's 3rd voyage had a "countermizzen" sail among its supplies, Lyon argues that the *Niña* had not 3 masts but 4, including a countermizzen just forward of the stern. Artist's illustration by Richard Schlecht, pp. 604-5.

[707] **José María Martínez-Hidalgo.** "Las naves de Colón y la polémica que no cesa," *Rev Gen Marina* (Madrid), 211 (1986): 477-95.

Reviews the various sources of information on CC's ships: the *Journal*; drawings, paintings, and letters representing the ships; archival notices; citations by historians and travelers; and early treatises on navigation and naval architecture.

Isabel Arenas Frutos

[708] **José M. Martínez-Hidalgo.** "Las naves de los cuatro viajes de Colón al nuevo mundo," *Temi Colombiani* (Genoa: ECIG, 1986), pp. 201-229.

Besides providing a convenient history of the reconstructions of CC's fleet of the 1st voyage (pp. 207-218), identifies the following ships as participating in the other voyages: (Voy 2) *Maríagalante* (capitana), officially called *Santa María; Niña* (officially *Santa Clara*), usually identified with the *Niña* of Voy 1, *San Juan, Cardera,* and *Gallega*; and, built in Isabela, *India* (officially *Santa Cruz*). (Voy 3) *Rábida, Garza,* and an unnamed ship were sent straight to Santo Domingo from Hierro in the Canaries; CC took the others to Trinidad and Paria, viz., *Santa María de Guía, Castilla, Gorda.* M.-H. accepts Barreiro Meiro's opinion that the caravels *Correo* and *Vaqueñad* (named by F. Columbus as being on the 3rd voyage) were the official names of *Gorda* and *Castilla* respectively. (Voy 4) *Capitana* (officially *Santa María); Bermuda* (officially *Santiago de Palos); Gallega*; and *Vizcaino.*

Discusses the movement of these ships during the voyages and supplies other interesting sidelights.

[709] **Carla Rahn Phillips.** "Sizes and Configurations of Spanish Ships in the Age of Discovery," *Columbus and His World* (Ft. Lauderdale FL: CCFL, 1987), pp. 69-98.

Discusses the meaning of the 5 measurements used to define Spanish ships in the Columbian period and thereafter. Analyzes the common formulas used to calculate ship tonnages based on those 5 measurements. Tables listing more than 100 real and ideal ships, small and large, accompany the article. The best estimates we have for CC's ships are examined in light of the analysis above. Develops a new set of estimates for CC's ships and a method for estimating the measurements of other historical ships.

2. Ships of the First Voyage

[710] **Spain.** *La nao "Santa María": memoria de la comisión arqueológica ejecutiva.* Madrid: El Progreso, 1892. 41 pp.

A detailed account of the Spanish construction of a full-scale model of the *Santa María* as part of the 4th centennial celebration.

[711] **Celso García de la Riega.** *La Gallega, nave capitana de Colón en el primer viaje de descubrimientos. Estudio histórico.* Pontevedra: Imp. de la Viuda de J. A. Antúnez, 1897. 199 pp.

Argues that Juan de la Cosa's *não,* the *Santa María,* was built in Pontevedro, Galicia.

[712] **Julio F. Guillen Tato.** *La carabela "Santa María," apuntes para su reconstitución.* Madrid: Ministerio de Marina, 1927. 235 pp.

Book I, pp. 9-60, debates the question whether the *SM* was a caravel or a nao, and concludes it was a caravel. Book II, pp. 65-155, discusses dimensions and form, fittings and sails, artillery, anchors & chains, etc., navigational instruments, and the crew's quarters. 9 appendices.

[713] **A. Stanley Riggs.** "*Santa María III*: The Twentieth-century Reproduction of Columbus's Flagship at the Seville Exhibition," *Art and Archaeology* (Wash DC), 28 (1929): 27-34.

A description of the Spanish government's replica, built in 1928.

Mug 85

[714] **G. S. Laird Clowes.** *Sailing Ships in the Science Museum.* London, 1932.

2: 18-19, a description of "the South Kensington model," a Spanish model of the *Santa María.*

[715] **Carlos Etayo Elizondo.** *La Santa María, la Niña, y la Pinta.* Pamplona: Isuna, 1962. 130 pp.

Reviews the history of ships and their construction in the era leading up to 1492; then describes three successive plans for reconstructing one or more of the ships (Spanish Commission, 1892; Julio Guillen; D'Albertis); then proposes specifications for a new reconstruction of all three ships.

[716] **John Frye.** *The Search for the Santa María.* New York: Dodd, Mead, 1973. 174 pp.

An account of Fred Dickson, Jr.'s attempt to ascertain whether a ship's keel in the sand of the barrier reef off Cape Haitien on the north coast of Haiti is the *Santa María*. Maps.

3. Ships of the Third Voyage

[717] **Roberto Barreiro-Meiro.** "Las naves del tercer viaje de Colón," *Rev Gen Marina* (Madrid), 178 (1970): 147-53.

Identifies the ships in CC's 3rd voyage as the não *Santa María de Guía* (flagship); *La Castilla, La Gorda, La Rábida, La Garza,* and another whose name is unknown. The first three CC took to Trinidad, where he discovered South America; the others he sent directly to Santo Domingo from the island Hierro in the Canaries. Chides Morison for failing to identify all these ships in the testimony of the *Pleitos* and for therefore being unable to identify them in *AOS* (#132).

Cf. Juan Gil, #362, and J.M. Martínez-Hidalgo, #708.

O. THE VESPUCCI QUESTION

[718] **Cesáreo Fernández Duro.** "Observaciones acerca de las cartas de Amerigo Vespucci," *Bol R Acad Hist* (Madrid), 8 (1886): 269-309.

Argues from style, content, and the things known about V. that this Florentine pilot, friend of CC, could not have written either (1) the letter addressed to Laurentio Petri Francisci de Medicis, recounting the "third voyage," first published in Paris by Jean Lambar, prior to the 1504 edition by Ottmar; or (2) the letter published by Hupfuff in 1505 recounting events on all "four voyages."

[719] **Juan Pérez de Guzman.** "Sobre el nombre de América y los demás que se dieron á las tierras occidentales descubiertas por Cristóbal Colón y los españoles," *El Centenario* (Madrid), 2 (1892): 249-69.

Reviews, among other things, what could be known about the naming of the New World for V. instead of CC.

[720] **Luigi Hugues,** "Amerigo Vespucci, notizie sommarie," *Raccolta*, V.2 (1894): 111-150.

An account of V's career, very sympathetic to his efforts and rejecting suggestions of duplicity on his part.

[721] **Alberto Magnaghi.** *Amerigo Vespucci: Studio Critico.* Rome: Arte Grafiche, 1924. 2 vols.

Setting aside the apocryphal works that brought V. into bad odor as a supposed thief of CC's claim to have discovered South America, M. strongly asserts, on the basis of the authentic documents, the Florentine navigator's claim to legitimate fame as a peer of CC and Magellan, both in his navigation and in his ability to conceive and project fruitful expeditions. Also, V's rank as a cartographer is very high.

[722] **Roberto Almagià.** "A proposito dei viaggi di Amerigo Vespucci," *Riv Geog* (Florence), 42 (1935): 49-56.

Cites testimony from *Los Pleitos* to establish that V. made a voyage to Cape St. Augustine, Brazil, for the King of Portugal in 1501-1502. This and the earlier voyage to Venezuela with Ojeda cannot be in doubt. Since V's skill at finding latitude is universally recognized, it cannot be claimed that he was not a mariner.

[723] **Giuseppe Caraci.** "I problemi vespucciani e i loro recenti studiosi," *Studi Colombiani* (Genoa: SAGA, 1952), 2: 495-551.

Traces the long controversy on (1) the authority of V's several voyages and (2) his reputed complicity in a movement to give the credit for CC's achievements to V. himself. A tradition of V as a malefactor begins in 18th c; Humboldt's *Examen critique* (1836-39), #123, draws together the known documentation on V. and places the events in context. Varnhagen describes 2 groups of sources: the printed and the MSS. He attributes the imputation of wrong to V. exclusively to the MS sources, which he considers apocryphal. Harrisse in turn contributes strongly to the rehabilitation of V., although he concedes in 1900 the doubtful character of V's supposed 1st voyage. Uzielli strongly advances the credit given to the MS letters as authentic. Vignaud's *Americ Vespuce* (Paris, 1917) supports V, but does little to refute the traditional rejection of the MSS sources. Finally in 1924 Magnaghi in *Amerigo Vespucci: Studio Critico* realizes a harmonious synthesis by establishing sound principles for evaluating and authenticating the sources. 3 subsequent scholars have made important contributions to V study: (a) J. F. Pohl (see Caraci's "Un nuovo libro sul V," *Riv Geog*, 54 [1947]: 117-22) establishes two voyages for V: 1499-1500 for Spain, and 1501-2 for Portugal. (b) Roberto Levillier sponsors V's achievements in *América la bien llamada* (Buenos Aires, 1948); not concerned, however, with textual authority or doubtful of *four* voyages. (c) T. O. Marcondes de Souza, *Amerigo V e sus viagens* (Soa Paolo, 1940) follows Magnaghi.

[724] **Paolo Revelli.** "Nuovo contributo di Roberto Levillier allo studio delle fonti piu antiche sui viaggi del Vespucci," *Studi Colombiani* (Genoa: SAGA, 1952), 2:649-71.

Defends Levillier against Caraci's attack (#723), which faults L. for continuing to accept all 4 Vespucci voyages as authentic despite rejection by those who approve Magnaghi's analysis of the sources. Cites and applauds L's 1951 work *Amerigo V: el Nuevo Mundo, cartas relativas a sus viajes y descubrimientos* (Buenos Aires: López) which Revelli perceives as authenticating Levillier's position.

[725] **Carlos Seco Serrano.** "Algunos datos definitivos sobre el viaje Hojeda-Vespucio," *Rev Indias* (Madrid), 15 (1955): 89-107.

Argues (a) that Hojeda in his 1499 voyage to South America set out from Puerto de Santa María on 18 Apr with one caravel, and was joined by another off north Africa at Cape Aguer; (b) that the two captains were Hojeda and Guevara; 4 pilots were named Juan Vizcaino, Juan Sánchez, Chamorro, and Juan López de Sevilla; and 20 crewmen were named, but not Vespucci; and (c) that the route proceeded from Santa María to Aguer and Torococa, and then to the islands of Lanzarote, Fuerteventura, Gran Canaria, Tenerife, and Gomera. Vespucci was *not* a captain on this voyage. Concludes that Vespucci was not on the 1499 voyage.

[726] **Carlos Sanz.** *El nombre América: libros y mapas que lo impusieron.* Madrid: Suarez, 1959. 244 pp.

Description and history of publications (1) of Waldseemüller's *Cosmographiae Introductio*, 1st ed. 1507, which on leaf 15 assigns the name "America" to the South American continent in tribute to Vespucci, the presumed discoverer, and (2) of the world map included between leaves 18 and 19, which bears the name "America."

[727] **Ramón Esquerra Abadía.** "Los primeros contactos entre Colón y Vespucio," *Rev Indias* (Madrid), 36 (1976): 19-47.

The amicable relationship between CC and V. could have begun as early as 1485, but is established by the evidence of joint dealings with CC's factor Berardi in the period 1493-95. Apparently the good relations were never broken, for CC wrote to Diego on 5 Feb 1504 of his liking for V.

[728] **Harold Jantz.** "Images of America in the German Renaissance," *First Images of America* (Berkeley: Univ. Cal Press, 1976), 1: 91-106.

Discusses several imaginative contributions of the German Renaissance that were taken into the total Renaissance image of America. Of special interest to Columbists is the account of the intricate humanistic process of determining the name of the new continent (pp. 96-100) constituting an amalgam of Vespucci's given name "Amerigo" with its rich potential for Greek puns, with the widespread occurrence of place names similar to "America" from Brazil to Venezuela, including Amaracao, Maraca, Marica, Maracaibo, Marahuaca, and Juan de la Cosa's "El Macareo."

[729] **Ilaria Luzzana Caraci.** "Punti di contatto e divergenze tra la storiografia vespucciana e quella colombiana," *Pres Ital Andalu II* (Bologna: Cappelli, 1986), pp. 143-55.

Reviews the curious controversy between supporters of CC and supporters of V., based on misunderstandings and confusion in the documents, a controversy that has quieted somewhat with the emerging fact that CC and V. were associates and good friends, and that V's supposed attempt (in the *Letter to Soderini*) to seize CC's fame is almost certainly apocryphal. Covers much of the ground treated more thoroughly in #730.

[730] **Ilaria Luzzana Caraci.** *Colombo e Amerigo Vespucci.* Genoa: Edizioni Culturali Internazionali Genova, 1987. 93 pp.

Monografie su temi Colombiani, no. 2. A review of V's career demonstrating that the supposed attempt by V. to seize CC's fame is not so and that V. and CC were friends and collaborators in the opening up of the newly found lands.

A struggle between CC's and V's supporters grew up after V's 1502 letter to Lorenzo di Pier Francesco dei Medici was converted by an unknown fabricator into two apocryphal Vespuccian letters, *Mundus Novus* and *Lettera al Soderini*, the second of which attempts to divert CC's credit to V. But the diversion of CC's fame that resulted from the name "America" had a certain justice in that it was originally applied only to South America, whose coasts V. not only explored extensively but which he—unlike CC—recognized as a continent not connected with the Orient.

Good bibliography of the CC—V. controversy of the 20th c.

VI

Bibliographies

A. GENERAL

[731] **Henry Harrisse.** *Bibliotheca Americana Vetustissima: A Description of Works Relating to America Published between the Years 1492 and 1551.* New York: G.P. Philes, 1866. 519 pp.

304 entries. Introduction, after a panegyric of the function of bibliography in helping organize the branches of knowledge, reviews previous bibliographies of Americana, and states H's own principles. The entries minutely describe and locate a copy of each work, mention the location of other copies, and comment on the work. A bibliographer's bibliography. Index by Carlos Sanz, next paragraph.

Updated several times by Carlos Sanz: 1. *El gran secreto de la carta de Colón y otras adiciones a la BAV* (1959), #100; 2. *Biblioteca Americana Vetustissima. Ultimas adiciones* (Madrid: Suarez, 1960), 3 vols., which updates Harrisse to 1551; 3. *Bibliotheca Americana Vetustissima; comentario crítico e índice general crónologico de los seis volúmenes que componen la obra* (Madrid: Suarez, 1960), 79 pp.

[732] **Giuseppi Fumagalli and Pietro Amat di S. Filippo.** *Bibliografia degli scritti italiani o stampati in Italia sopra Cristoforo Colombo, la scoperta del nuovo mondo e i viaggi degli Italiani in America. Raccolta,* VI (Rome: Ministero della Pubblica Istruzione, 1893). 217 pp.

Sections on CC's precursors, items 1-18; on CC, items 19-696, divided into CC's writings (19-81); writings about CC (82-546); allusions to CC & his discovery 1491-1550 (547-685); and editions after 1550 of early documents on the discovery (686-96).

A further section (697-1400) lists (1) works on America by Italian authors, and Italian translations of works on America, including archaeological and lingusitic studies; and (2) narratives by Italian travelers in America.

A ground-breaking bibliography, with items numbered in a single series (a practice not adopted in the USA until many years later). Items carefully classified and mostly annotated.

[733] **Real Academia de la Historia.** *Bibliografía Colombina. Enumeración de libros y documentos concernientes á Cristóbal Colón y sus viajes.* Madrid: Fontanet, 1892. 680 pp.

8 sections: (1) documents related to CC & his descendants; (2) CC's writings; editions of them; works discussing his writings and the printed editions; (3) works about CC (including general biographies, limited biographical studies, and Columbiana); (4) histories that touch on CC; (5) bibliographies and other reference works; (6) literary works about CC; (7) artistic works about CC; (8) works about the celebration of the 4th centennial.

As the chief 19th-c Spanish contribution to the enumerative bibliography of CC, this must of course be included. For a wry appraisal, see H. Harrisse, *CC et les Académiciens Espagnols* (Paris: Welter, 1894). H. condemns the work for lame planning, as seen in the overlapping categories; for impossibly vast categories (e.g. section 4); for the absurd exclusion of journal articles, which are the lifeblood of the historical discipline; and for various other blunders.

[734] "Travaux de la commission des grands voyages et des grandes découvertes," *Bulletin of the International Commission of Historical Sciences,* 7 (1936): 363-445.

See sections 9 and 10, pp. 409-429, for items on CC.

[735] **Donald H. Mugridge.** *Christopher Columbus: A selected List of Books and Articles by American Authors or Published in America, 1892-1950.* Washington: Library of Congress, 1950.

An annotated typescript, 110 items. "America" means USA. "American" means "born in the US or its dependencies."

[736] **Luis Florén Lozano.** "Bibliografía Colombina," *El Faro á Colón* (Ciudad Trujillo), no. 8 (1953), pp. 109-129; no. 9 (1953), pp. 163-75.

A list of the books by or about CC in the libraries of the Dominican Republic. 211 entries.

[737] **Anthony Tudisco.** "América en la literatura española del siglo xviii," *Anuario Estud Amer* (Seville), 11 (1954): 565-85.

A bibliography of 18th-c Spanish writings on CC and the discovery.

Martin Torodash

[738] "America en la bibliografía Española (reseñas informativas)," *Hist y Bibl Amer* (Seville), 1 (1955)-33 (1988).

An annual annotated bibliography on Americanist studies, editions, archival guides, and bibliographies in Spanish publications. Supplement to *Anuario de Estudios Americanos* (Seville) beginning 12 (1955).

[739] **José Alberto Aboal Amaro.** *Catalogo sistemático de la Biblioteca Colombina de Montevideo, República Oriental del Uruguay.* Montevideo: Ediciones de la Biblioteca, 1966. 144 pp.

A catalogue of the monumental collection assembled by Aboal Amaro at the Biblioteca Colombina in Montevideo, and acquired by SUNY – Stony Brook in 1966. Most of the books listed are now in the general collection of the library at Stony Brook; the rare ones are in the special collections.

Sections: CC, general and particular studies; fatherland; family; Genoa in the epoch of CC; ships of the time; CC in Portugal; CC in Spain; Pontevedra; voyages: routes, accounts, and letters; burial; los *Pleitos*; monuments; portraits & medals; lighthouse; the signature; codices; 4th centennial celebration; CC in art; canonization; prediscovery; discoveries in general and particular; general studies of the discovery of America; particular studies of it; legislation; exploration and conquest; histories of the New World and of the Indies; raccoltas, annuals, bulletins, memorials, etc.; bibliographic indices; Toscanelli, Vespucci and the name America; navigation to the West Indies; the black legend; 18th-c philosophical judgments on the discovery; teaching of Christianity; cartography; geography; Portuguese enterprises; Columbian bibliography; bibliographies of American studies; maritime bibliography; book notices; book reviews; meetings; chronicles of the discovery; biographies and bio-bibliographical studies; journeys and travelers; studies of American themes.

[740] **Genoa. Berio Civic Library.** *Catalogo della raccolta colombiana,* 2nd ed. Maria Tina Pareto Melis and Giacomina Calcagno, eds. Boston: G. K. Hall, 1963. 151 pp.

3157 entries. Photographic reproduction of catalogue cards. Project directed by G. Piersantelli. 1st edition, ed. Luigi Augusto Cervetto (Genoa: Pagano, 1906). 126 pp.

The Berio collection is probably the best single collection of Italian items on CC, and contains a very good sampling of items from other countries. This catalogue is supplemented at the library by a regularly updated catalogue of addenda.

[741] **Barbara G. Cox,** ed. *HAPI: Hispanic American Periodicals Index.* Los Angeles: UCLA Latin American Center Pubs, 1975-1986.

After 1976, editor's name is Barbara G. Volk.

[742] **Grazia Galliano.** "Guida alle opere di interesse Colombiane . . . [in] Genoa," *Atti III Conv Internaz Stud Col* (Genoa: 1979), pp. 713-722.

Surveys the books and other documents with a primary or partial emphasis on geography to be found in the Berio Civic Library, the Biblioteca Universitaria, the Biblioteca dell'Associazione Italiana di Studi Americanisti, and the Biblioteca dell'Istituto di Scienze Geografiche della Facultà di Magistero.

[743] **Gabriella De Paoli, Maria Giuseppina Lucia, and Graziella Galliano.** *Contributi alla bibliografia colombiana.* Genoa: Tilgher, 1980. 140 pp.

Contains 2 bibliographies, each preceded by an essay: De Paoli and Lucia's "Per una bibliografia colombiana (1917-77)" and Galliano's "Guida alle opere di interesse colombiano conservate presso le principali biblioteche di Genova. Saggio bibliografico." (See #742). Galliano's bibliography very helpfully records the Genoese library call numbers for each item.

All the items dated 1880 and afterward have been incorporated into Simonetta Conti's 1985 bibliography, #745, but frequently Conti has omitted significant bibliographical details.

[744] **Gabriella Boscolo.** "Saggio de bibliografia colombiana," *Saggi e Documenti* (Genoa), 2, tome 2 (1981): 400-59.

Printed books and articles on CC in the libraries of Madrid and Barcelona, and those in the Biblioteca Nazionale in Rome. No indication of page numbers for articles, and many lacunae in such bibliographical information as is recorded.

[745] **Simonetta Conti.** *Un secolo di bibliografia colombiana* (1880-1985). Genoa: Cassa di Risparmio, 1986. 360 pp.

3271 entries. An indispensable beginning in drawing together CC studies. Somewhat difficult to use because of limited index and absence of annotations. Useful list of journals surveyed (pp. 335-48).

[746] **Anna Maria Salone.** *Opere colombiane della biblioteca universitaria di Genova.* Genoa: A. Compagna, 1987. 187 pp.

1458 items, chronologically entered, with call numbers. Divisions: (1) MSS & autographs, chronologically listed; (2) printed works chronologically listed; (3) undated printed works.

B. SPECIALIZED

[747] **Servando Arboli Faraudo.** *Biblioteca Colombina: Catalogo de sus libros impresos de la santa metropolitana y patriarcal iglesia de Sevilla.* Seville: E. Rasco, 1888-94. 7 vols.

A description of each book in what is left of Ferdinand C's library, alphabetized by author or, if anonymous, by owner if known. Following each entry is transcribed FC's note in the book, often indicating the price, the date acquired, and his own serial number. Editor gives extensive information on the very important books, e.g., CC's copy of D'Ailly's *Imago Mundi.*

[748] **El Conde de Viñaza.** "Bibliografía española de lenguas indígenas de América," *El Centenario* (Madrid), 2 (1892), 57-67.

Prints the introduction to the author's 1892 bibliography by this title (Madrid: Biblioteca Nacional, 1892).

[749] **Giuseppe Fumagalli.** "Bibliografia delle opere concernenti Paolo Toscanelli ed Amerigo Vespucci," in *Vita di Amerigo Vespucci* (Florence: Auspice il Comune, 1898), pp. 99-131.

A sequential record of 57 documents referring to T from 1533 to 1898, and a sequential record of 280 docs. referring to V from the 16th c to 1898.

[750] **Henry Vignaud and G. Uzielli.** *Bibliografia della polemica concernente Paolo Toscanelli e Cristoforo Colombo.* Naples: Tocco-Salvietti, 1905.

169 items. Introduction by Uzielli describes the issues and incorporates 2 documents he has discovered, i.e., a 1454 conversation between T and the Portuguese ambassadors and a 1494 letter of the Duke of Ferrara seeking information about Toscanelli.

[751] **Archer M. Huntington.** *Catalogue of the Library of Ferdinand Columbus: Reproduced in facsimile from the Unique Manuscript in the Columbine Library of Seville.* New York: Privately printed, 1905. 260 pp.

Rpt. New York: Kraus, 1967. Facsimile reprint of the manuscript in the Colombina Library in Seville. Minute description of each book, in FC's hand, detailing cost and place and date of purchase. 4231 numbered items.

[752] **Paolo Revelli.** "Manoscritti relativi alle terre d'America conservati nelle biblioteche e negli archivi d'Italia," *Terre d'America e archivi d'Italia* (Rome: Treves, 1926), pp. 59-173.

Lists and describes early cartographical documents, accounts of voyages, and descriptions of lands in the western hemisphere. Surveys the libraries and archives of the whole country.

[753] **Emilio Rodríguez Demorizi.** *Colón en la Española: itinerario y bibliografía.* Ciudad Trujillo: Academia Dominicana de la Historia, 1942. 42 pp.

1. Lists memorable dates of CC and his descendants relating to Santo Domingo and those later dates relating to CC homages, heraldry, and monuments, including the Columbus lighthouse. 2. Lists the Dominican works relating to CC himself, his remains, and the CC statue, mausoleum, and lighthouse.

[754] **Carlos Sanz.** *Bibliografía general de la Carta de Colón.* Madrid: Suárez, 1958. 305 pp.

Lists the printed translations and editions of CC's *Letter* of 4 Mar 1493 informing the monarchs of his discovery. Elaborate and informative annotations. 1st entry, the Latin Rome ed. of May 1493.

[755] **Enrique Bayerri Bertomeu.** *Colón tal cual fué; los problemas de la nacionalidad y de la personalidad de Colón y su resolución mas justificada.* Barcelona: Porter-Libros, 1961. 802 pp.

A generously annotated bibliography of items dealing with the two issues named in the title. The author's commentary, however, moves to the curious conclusion that CC was a native of an island named Genova in a Spanish river.

[756] **Simonetta Conti.** "Bibliografia delle edizioni del *"Diario de a bordo"* o *"Libro de la primera navegacion,"* in *Il Giornale di Bordo, Libro della Prima Navigazione e Scoperta delle Indie,* ed P.E. Taviani and Consuelo Varela. (Rome: I.P.Z.S, 1988), II.453-69.

Nuova Raccolta Colombiana, Vol. 1. 85 entries, 1825-1987. Many items annotated.

VII

Columbus Scholarship:
Questions, Methods, Achievement

[757] **Henry Harrisse.** *D. Fernando Colón, historiador de su padre. Ensayo critico.* Seville: Sociedad de Bibliófiles de Sevilla, 1871.

French edition, *Fernand Colomb: sa vie, ses oeuvres; essai critique par l'auteur de la Bibliotheca Americana Vetustissima* (Paris: Tross, 1872), 230 pp. Begins the attack on the *Historie* as inauthentic. A study of the life of CC's son Ferdinand and of the *Historie,* a biography of CC attributed to Ferdinand. Concludes that F cannot have been its author. Begins the long series of attacks on the biographies of Ferdinand and Las Casas extending to the middle of the 20th century.

[758] **Eugene Gelcich.** *La Scoperta d'America e Cristoforo Colombo nella letteratura moderna.* Gorizia, Italy: Paternolli, 1890. 151 pp.

An attempt to place Columbus study within the proper historical and critical contexts. Chapters: 1. CC's precursors. 2. Nautical science in the Age of Discovery. 3. CC's nautical expertise. 4. CC and contemporary criticism (late 19th century). 5. Critical issues: (a) the authenticity of Ferdinand C's *Historie;* (b) the date of CC's birth; (c) determining the site of the 1492 landfall and the elements involved in studying the problem, viz., magnetic declination, ocean currents, the length of CC's mile. 6. The suit of Diego C. and his descendants against the crown.

[759] **Marcelino Menéndez Pelayo.** "De los historiadores de Colón," *El Centenario* (Madrid), 2 (1892), 433-53; 3 (1893): 51-7.

Rpt. in *Estudios y discursos de crítica histórica y literaria,* ed. D. E. Sánchez Reyes (Santander: Artes Gráficas, 1942), 12: 69-122.

Part 1 is a vigorous account of the primary sources of CC biography, serving as a prelude to a review of modern CC biography, which appears in part 1. This review touches on Robertson (1777), #118; Muñoz (1793), #121; Navarrete (1825), #2; Irving (1828), #122; Humboldt (1836), #123; Roselly de Lorgues (1856), #655; Harrisse (*BAV,* 1866, #731; Life, 1884, #125); and

Fernández Duro (various works). All this leads up to a review of J. M. Asensio's *Cristóbal Colón* (1888), #126, which is the declared purpose of this article. Menéndez Pelayo does not hesitate to cite extensive weaknesses in Asensio's life of CC, but in general praises it. For a more acerbic opinion, see Harrisse's review, cited in annotation to #126.

[760] **Cornelio Desimoni.** "Quistioni [*sic*] Colombiane," *Raccolta*, III.3 (1892): 7-126.

States and attempts to resolve (or to acknowledge the moot status of) the many issues facing CC scholarship at the time of the 4th centennial.

[761] **Roberto Almagià.** "Pietro d'Ailly e Cristoforo Colombo," *Riv Geog* (Florence), 38 (1931): 166-69.

Review of E. Buron's *Imago Mundi de Pierre d'Ailly* (Paris: Maisonneuve, 1930), #65, which includes CC's marginal notes. Compliments Buron for rejecting Vignaud's hypothesis that CC read d'Ailly only *after* the first voyage, but criticizes him for placing CC's reading of d'Ailly prior to 1483, for dodging the question of whether the notes are CC's or Bartholomew C's, and for accepting Roncière's thesis that the Bibliothèque Nationale map was drawn by CC.

[762] **Rómulo D. Carbia.** *La nueva historia del descubrimiento de América.* Buenos Aires: Coni, 1936. 151 pp.

Revives the attack, begun by Harrisse in *D. Fernando Colomb, historiador de su padre* (see above, #757), on the biographies of CC by Ferdinand Columbus and Las Casas. Treats Las Casas with especial contempt and acerbity. Believes Ferdinand's *Historie* is an invention by Las Casas to help the Colóns in their suit against the crown. Calls for re-evaluation of existing sources and correction of printed versions by returning to the manuscripts. A salutary, though not always just (i.e., to Las Casas), reminder that good scholarship depends on accurate texts. See #763.

[763] **Roberto Almagià.** "La nuova storia della scoperta dell'América," *Boll Soc Geog Ital* (Rome), 74 (1937): 171-82.

Carbia's book by this title, #762, has two defects: (1) he does not take the historical framework sufficiently into account; (2) he is therefore unable to judge facts and documents in the light of geographical and scientific knowledge of CC's time, so different from today. Almagià calls for critical editions of Las Casas' *Historia de las Indias,* of Ferdinand C's *Historie,* and of *Los Pleitos.*

Rina Ferrarelli Provost

[764] **Rómulo D. Carbia.** *La investigación científica y el descubrimiento de América.* in *Historia de la Nación Argentina.* Buenos Aires: Privately published, 1937. 73 pp.

Continues the attack begun half a century before by Harrisse (in #757) on what Carbia calls "the traditional thesis" about CC and his Enterprise.

Calls into question and/or denounces as fraudulent the *Capitulations of 1492,* the sovereigns' letter to the Grand Khan, CC's passport, Las Casas' account of the enterprise, the letter and map attributed to Toscanelli, and the annotations attributed to CC in the books in Ferdinand Columbus's library.

Since 1952, this issue has been largely ignored as spurious, and both the CC documents and Las Casas' account have been accepted as fundamentally sound.

[765] **Charles E. Nowell.** "The Columbus Question: A Survey of Recent Literature and Present Opinion," *Amer Hist Rev,* 44 (1939): 802-822.

A trenchant, carefully documented essay reviewing Columbus studies of the early twentieth century.

[766] **I. Oreste Bignardelli.** "Circa la necessità d'una nuova rivaluazione delle fonti della storia colombiana e della scoperta," *Studi Colombiani* (Genoa: SAGA, 1952), 2: 443-49.

An appeal for scholarly rededication to evaluating and properly editing the basic sources of CC study, esp. Ferdinand's *Historie,* Las Casas' *Historia,* and the *Pleitos.* Calls for an end to the tendentious struggle between the supporters of Harrisse and de Lollis over the value and authenticity of Ferdinand C's *Historie,* which in 1951 was still animated and still was blocking a scholarly disposition of fundamental textual problems.

[767] **Ettore Rossi.** "Scritti turchi su Cristoforo Colombo e la scoperta dell'America," *Studi Colombiani* (Genoa: SAGA, 1952), 2: 563-66.

Reviews Turkish studies on CC, among which the only significant item is the discovery in 1929 of the Piri Re'is map. Includes a bibliography of 9 reports and studies on the map. Concludes with an account of a curious Turkish tradition reported in Ibrahim Hakki, *Topkapi Sarayinda deri üzerine yapilmis eski harita lar [Ancient parchment map of the Topkapi Palace]* (Istanbul, 1936), pp. 93-94. The tradition holds that CC appeared in Constantinople at some time after 1481 and sought the Sultan's aid for a voyage to "discover the new world" but was successfully opposed by learned Mussulmans of the court.

[768] **Marcel Battailon.** "Historiografía oficial de Colón de Pedro Martír a Oviedo y Gómara," *Imago Mundi* (Buenos Aires), 1:5 (1954), pp. 23-39.

Examines the assumptions and inclinations that led Gómara, self-appointed "official historiographer of the Indies," to reject CC's aim to reach Asia and to accept the "unknown pilot" story, whereas Oviedo, the truly official historiographer of the Indies, considered the pilot story unworthy of credit and simply ignored the Asian aim even though Peter Martyr, namer of the "New World," had labeled CC's aim to reach Asia a "natural" aim.

[769] **Martin Torodash.** "Columbus Historiography since 1939," *Hisp Amer Hist Rev*, 46 (1966): 409-428.

Continues the historiography of Columbus studies where Nowell leaves off (#765), and brings it up to 1966.

[770] **Ilaria Luzzana Caraci.** "Nuova luce intorno al problema delle 'Historie' colombiane," *Boll Soc Geog Ital* (Rome), 5 (1976): 503-512.

Rumeu de Armas, in his *Hernando Colón* (Madrid: Cultura Hispánica, 1974), #623, has shown the existence of an anonymous biography alongside FC's original text, as well as later interpolations and augmentations.

However, to resolve all the questions thus raised, in order to ascertain the origins and dimensions of the cited augmentations and interpolations, it would be neccesary to broaden the critical study to the various parts that make up the *Historie*.

[771] **Alberto Boscolo.** "Ricerche su Cristoforo Colombo e sulla sua epoca," *Atti III Conv Internaz Stud Col* (Genoa: CIC, 1979), pp. 51-58.

Much still remains to be done in the archives of southern Spain, especially the "Archivos de Protocolos" in Seville and Córdova and the Municipal Archive of Málaga. Research can go in 3 directions: the ambience of CC's life; his family; and the Genoese active in the Iberian peninsula before and after the discovery.

[772] **G. DePaoli and M. G. Lucia.** "Saggio di bibliografia colombiana (1951-77)," *Atti III Conv Internaz Stud Col* (Genoa: CIC, 1979), pp. 391-402.

Reviews studies dealing with the Columbian issues active from 1951 to 1977: CC's language, culture, technical nautical knowledge, possible precursors, voyages, esp. the 1st, and the landfall.

[773] **Ramón Ezquerra Abadia.** "Medio siglo de estudios Americanos," *Anuario Estud Amer* (Seville), 38 (1981): 1-24.

A survey of Americanist studies, 1930-80.

[774] **Manuel Romero Tallafigo.** "El archivo general de las Indias: acceso a las fuentes documentales sobre Andalucía y América en el siglo XVI," *Andalucía y América en el Siglo XVI. Actas de las II Jornadas de Andalucía y América* (Seville: Escuela de Estudios Hispano-Americanos, 1983), 1: 455-84.

An essay describing the formation and organization of the Archive of the Indies in the Casa de Contratación in Seville. Fundamental for the beginner in using this rich and complex archive.

[775] **Ilaria Luzzana Caraci.** "Genova e Colombo," *Scritti Geografici di Interesse Ligure* (Genoa: Ist. di Scienze Geografiche, 1984), pp. 227-55.

Reviews the growth of the idea that CC was not Genoese, and the growth of Columbian historiography, especially as pursued in Genoa, beginning with Spotorno's *Della origine e della patria di CC* (Genoa: Frugani, 1819), with special attention to the genesis of the 1892-96 *Raccolta*.

[776] **Ilaria Luzzana Caraci.** "Gli studi colombiani e il quinto centenario della scoperta dell'America," *Riv Geog* (Florence), 91 (1984): 111-125.

Points out Italian achievements in CC research, dominated by geographers. Turns to current problems looming as the 5th centennial approaches, particularly the conceptual genesis of the discovery, a problem closely linked to the authenticity of documents and with the analysis of CC's autographs, especially his letters and his marginal notes.

[777] **Simonetta Conti.** "Orientamenti bibliografici colombiani," *Columbeis I.* (Genoa: IFCM, 1986), pp. 77-91.

A brief survey of CC scholarship of the past century, with chief emphasis on the vigorous activity during the 70's and early 80's.

[778] **Gaetano Ferro.** "Storia delle esplorazioni e geografia: l'esempio delle opere di Paolo Emilio Taviani," *Temi Colombiani* (Genoa: ECIG, 1987), pp. 175-82.

Distinguishes between mere accidental discovery and geographical exploration deriving from careful reflection on the ambient human circumstances within which certain phenomena of the earth's crust have come to light. The latter advances theoretical comprehension of the planet.

Proceeds to the observation that current advances in historical knowledge presuppose cooperation among disciplines, embracing geographical and linguistic as well as geographical assumptions and experience, in addressing the source of historical knowledge, i.e, the documents. Cites P.E. Taviani's *CC: La genesi della grande scoperta*, #441, and *I Viaggi di Colombo*, #252, as examples of historical study drawing the disciplines together productively.

[779] **Jacques Heers.** "Le projet de Christophe Colomb," *Columbeis I* (Genoa: IFCM, 1986), pp. 7-26.

The background and sources of CC's Enterprise of the Indies.

[780] **Demetrio Ramos Pérez.** "El inicio de la historiografía americanista y el lugar donde se llevo a cabo: La datación del comienzo de las 'Decadas'de Pedro Martír de Anglería," *Temi Colombiani* (Genoa: ECIG, 1986), pp. 267-85.

Dates the composition of PM's first *Decade* in the period 1494-97.

Index of Authors and Editors

Italicized entry identifies an anonymous collection or reference tool

Index of Persons and Places

Place names are in italics

213

Index of Topics